PRAISE FOR THE ULTIMATE INTERPLANETARY TRAVEL GUIDE

"Like any good travel book, Jim Bell's guide gives you some good advice on what to bring and what to wear. Of course, we don't leave for a few hundred years. For now though, check out these images—the pictures alone will make you want to start packing!"

—Bill Nye, CEO, The Planetary Society

"This guide to our amazing solar system is loaded with approachable science, thought-provoking challenges and beautiful imagery to transport you there. This book belongs on every big dreamer's night stand!"

—Scott Parazynski, MD, Astronaut, Tech CEO, and Author of *The Sky Below*

"Want to hotwing on Venus or enjoy fine dining on Phobos? Then Jim Bell's is the essential guidebook to have in your spacesuit backpack. Where to go, what to see, how to pack, you'll be all ready to board your rocket when the year 2218 rolls around. Until we acquire the real Hitchhiker's Guide to the Galaxy, this book is the next best thing to reveal our solar system's hottest—and coldest—spots."

—Jon Lomberg, Space Artist and Design Director for the Voyager Golden Record

"The Ultimate Interplanetary Travel Guide is a fanciful vision of future hotels at lunar poles and cable cars up the volcanoes of Mars. It's both a fun ride and a fact-filled journey through the many worlds of the solar system—and beyond!"

—Emily Lakdawalla, Senior Editor, The Planetary Society

PRAISE FOR JIM BELL

THE SPACE BOOK: "An extraordinary compilation of all that matters, ever mattered, or ever will matter in the universe. While it's a stunning source of information, Jim Bell's *The Space Book* is nonetheless a delight to browse and even more fun to read."

–Neil deGrasse Tyson, Astrophysicist and Author of *Space Chronicles: Facing the Ultimate Frontier*

MARS 3-D: "Jim Bell takes us on an extraordinary journey across often mysterious, sometimes perilous, and always fascinating Martian terrain. A must-read for anyone who's ever dreamed of exploring the Red Planet."

–Buzz Aldrin, Apollo 11 Astronaut

OTHER BOOKS BY JIM BELL

POSTCARDS FROM MARS ✶ MARS 3-D ✶ MOON 3-D
THE SPACE BOOK ✶ THE INTERSTELLAR AGE

STERLING
New York

An Imprint of Sterling Publishing Co., Inc.
1166 Avenue of the Americas
New York, NY 10036

ISBN 978-1-4549-2568-2

Distributed in Canada by Sterling Publishing Co., Inc.
c/o Canadian Manda Group, 664 Annette Street
Toronto, Ontario M6S 2C8, Canada
Distributed in the United Kingdom by GMC Distribution Services
Castle Place, 166 High Street, Lewes, East Sussex BN7 1XU, England
Distributed in Australia by NewSouth Books
45 Beach Street, Coogee NSW 2034, Australia

For information about custom editions, special sales, and premium and corporate purchases, please contact Sterling Special Sales at 800-805-5489 or specialsales@sterlingpublishing.com.

Manufactured in China

2 4 6 8 10 9 7 5 3 1

sterlingpublishing.com

For image credits, see page 146

Cover design by David Ter-Avanesyan
Interior design by Gavin Motnyk
Interior layout by Ashley Prine

A FUTURISTIC JOURNEY THROUGH THE COSMOS

THE ULTIMATE INTERPLANETARY TRAVEL GUIDE

✳

JIM BELL

STERLING
New York

CONTENTS

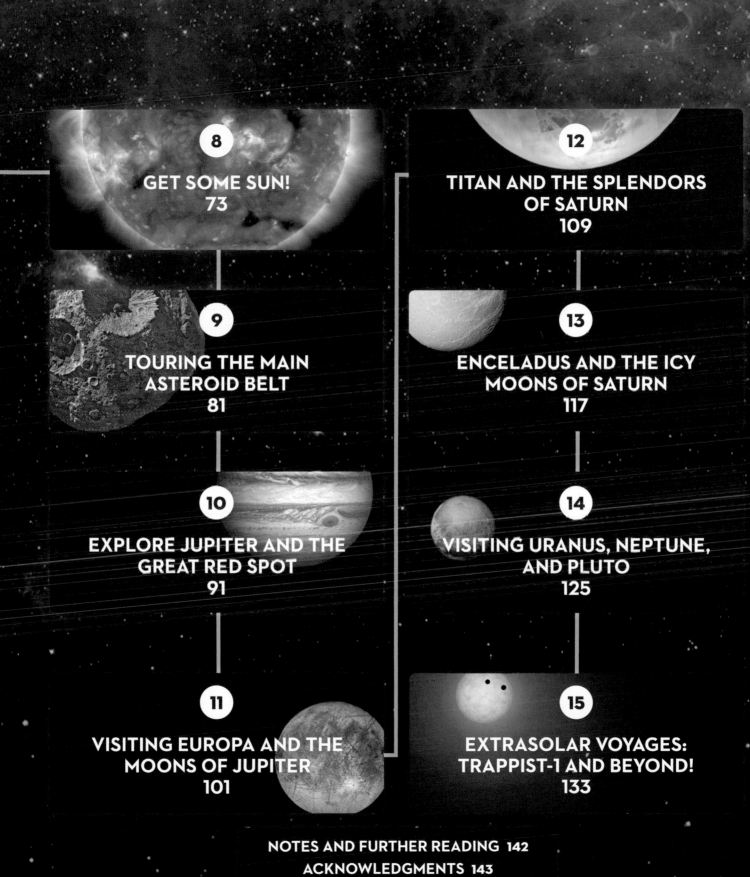

AUTHOR'S NOTE

WE ARE LIVING IN WHAT HISTORIANS WILL surely one day call the golden age of solar system exploration. That is, since the start of the space age in the late 1950s, humans have launched robots to fly by, orbit, land, or rove on nearly all the different kinds of worlds in our solar system. We've learned an enormous amount about the worlds around us and—just as importantly—about how special and unique (at least in our solar system) our home planet really is. The most daring exploits of human space exploration happened nearly 50 years ago, however, and today we are at a crossroads. After decades of being limited to low Earth orbit missions, what is the future of human space exploration?

I'm an astronomer and planetary scientist and have been actively involved in NASA's robotic solar-system-exploration missions for more than 20 years. I've seen many worlds go from mere points of light to exotic destinations I yearn to visit. And I've seen firsthand NASA and other space agencies design, build, and fly orbiters, landers, and rovers to many of these exotic worlds, paving the way for eventual human exploration. Because of this amazing progress during just our first 50 years or so as a space-faring species, I believe that it is inevitable that humanity will one day become a multi-planet—and multi-star-system—species.

Some of what will motivate us will be the thrill of scientific discovery and exploration of the unknown. But I believe that we will also be motivated by our spirit of wanderlust, by our strong desire to live lives full of new experiences and challenges—the same motivation that drives the enormous tourism industry here on Earth. Our planet offers tourist experiences of all kinds, which has turned out to be a successful business model for countless companies worldwide. In my opinion, that

model—buoyed by the development of the needed technologies over time—will eventually work for space-based tourism as well. That simple assumption is really the basis for this book.

Using reasonable extrapolations from current technologies and the scientific, environmental, and engineering knowledge needed to get to and visit the worlds around us, I've fancifully set the text 200 years in the future, as a travel guide for people contemplating a getaway vacation to space. Based on what we've learned about those worlds in the past half century, I've tried to keep my assumptions about the future of space tourism realistic (no transporters or warp drives needed!). However, many of the realities, even 200 years from now, are still likely to be sobering. The solar system is *big*, and even with reasonably assumed advanced-propulsion technologies, we are still likely to measure travel times to solar system destinations in weeks at best—and more likely in months or even years.

I've also chosen to be somewhat selective about future tourist destinations, partly because of practical limits on the length of the book, but also because I've tried to focus on what I believe will be the "must see" destinations of the solar system. Many of these places have recently been featured in a beautiful series of NASA-commissioned space travel posters, which have partly inspired my writing. Eight of those posters can be found in the back of this book as bonus material that you can remove and post on your walls. The potential of future space exploration and tourism across the solar system and beyond has inspired not only artists, but also poets, novelists, musicians, and filmmakers. Hopefully you will be inspired too, and will enjoy visiting, in your mind's eye at least, some of the truly grand spectacles on the worlds around us. Just like Earth, the rest of the solar system has much to offer those interested in geology, meteorology, photography, hiking, climbing, extreme sports, outdoor adventure, and even the history of science and exploration.

We've already seen the first few examples of space tourism. For example, the Russian space program has been ferrying "tourists" to the International Space Station since 2001, when American businessman and millionaire Dennis Tito booked a ride on a Soyuz spacecraft. Since then various other fantastically rich and/or influential individuals have paid for rides beyond Earth. So-called "newspace" companies have recently formed and are planning to offer people high-altitude balloon trips to the edge of space and suborbital spacecraft rides. Virgin Galactic, for example, has already sold more than 750 prepaid tickets for the latter, at $250,000 apiece. And while neither NASA nor any other national space agencies have yet planned specific human missions to Mars, several start-ups, including SpaceX and Mars One, have promoted trips to the Red Planet (some of them one-way) for anyone who can purchase a ticket. Public reaction has generally been positive but a little skeptical, since space travel is not yet an affordable or practical option for the next family vacation.

That could all change, however, over the next 50 to 100 years, as travel to space, including deep-space destinations, potentially becomes as routine as airline travel is today. The pieces are already falling into place: special government programs exist to enhance the capabilities of the commercial space industry (just as the government did back in the 1920s for the commercial aviation industry); entrepreneurial start-ups are focusing on human space travel for scientific, exploration, and tourism purposes; and intergovernmental discussions are starting about updating or otherwise amending international treaties related to the conduct of business in space and on other worlds.

If you had told someone in 1918 that, by the end of the century, it would be possible for an average middle-class family to travel almost anywhere on Earth in two or three airline flights, that person probably would have

MARS EXPLORERS WANTED

Wouldn't it be fun to go canyon climbing on Mars?

thought you were crazy. You're entitled, then, to think it's crazy for me to claim in 2018 that, by the end of this century, it may be possible to travel anywhere in the *solar system* in two or three spaceliner flights. Regardless of the timing, I believe that solar system tourist travel is coming eventually, and that the sheer number of off-world destinations and opportunities for spectacular sightseeing, unique adventures, and personal education and growth is going to expand dramatically, one day spanning the entire solar system and beyond. Fly safe, and be sure to book your seats in advance!

—Jim Bell
Mesa, Arizona

PREPARING FOR YOUR INTERPLANETARY ADVENTURE

So you've decided—finally—to take that long-awaited trip off planet. Well, you've come to the right place. Welcome to the 2218 edition of *The Ultimate Interplanetary Travel Guide*! We've pulled together some of the best tips, tricks, and must-see destinations to help you put together an amazing adventure. Whether you're planning a quick trip to the Moon or a more leisurely many-year cruise out to Pluto, you'll find a treasure trove of useful historical and scientific facts, figures, and insider information to make your voyage the trip of a lifetime.

Our solar system—and beyond—is filled with wondrous and spectacular natural and human-made destinations for you to explore. Indeed, your choices are almost limitless, and your adventure can be short or long, completely arranged for luxury on well-traveled paths, or individually tailored to let you tick off those out-of-the-way "bucket list" places you've always wanted to see. Singles, couples, families—there are so many choices for you all.

This guide is organized approximately by the duration of your trip, starting from close-to-Earth destinations in and around the inner solar system, moving out to the asteroid belt and the outer solar system, and even including a highly anticipated new opportunity to visit some Earthlike planets around a relatively nearby star called TRAPPIST-1. Our staff of professional astronomers, planetary scientists, astrobiologists, artists, and space technology and rocketry engineers have compiled an enormous amount of information, including some spectacular photos and inspirational posters that can help you plan the perfect trip. Want to climb the tallest volcano in the solar system? Dive off the tallest ice cliff? Barrel ride down a methane waterfall? Skim through a storm three times the size of Earth? Have a romantic dinner surrounded by a billion shimmering house-size crystals of ice? All this and more is possible, and the details you need are here.

In addition to describing many of the highlights of these destinations, we provide you with some exciting scientific facts and history on space exploration that can really enrich your trip. Who were the first explorers—human or robotic—to discover and chart these destinations, and what did we learn from those early missions? Why are some of these places designated as historic sites, as nature preserves, or as scientific research stations? Which of the many companies now working to mine natural resources like water and precious metals from the Moon, Mars, asteroids, and comets offer the best tours of their facilities? And what are some of your best options for accommodations, recreation, and dining? We've got the answers!

First, though, let's go over some basic logistics and important pre-trip planning you'll need to do to make your adventure successful.

WHAT TO PACK

Back in the early twentieth century, the pioneering French aviator Antoine de Saint-Exupéry summarized it best: "He who would travel happily must travel light." We strongly advise that you take this advice to heart when planning your upcoming interplanetary voyage.

Suit up, ship out, and explore the solar system.

While travel among the planets, moons, asteroids, and comets in our solar system is now routine compared to back in the early space age, it's still a challenge to bring lots of food, clothing, electronics, or other supplies along with you. Seasoned solar system travelers know, for example, that many spaceline carriers, public and private, charge extra luggage fees for mass or volume above certain limits, reflecting the realities of launch costs and cargo-hold sizes. So pack light and focus on the essentials:

Clothing Consider light and informal clothes for comfortable zero-gravity (zero-g) cruising between your destinations, good gripping socks (many still prefer the old Velcro® variety) for standard low gravity–simulated tunnels and corridors in some of the larger vessels, and perhaps shorts, T-shirts, and comfortable sneakers if

UNDERSTANDING GRAVITY

Everything in the universe attracts everything else, and the intensity of that attraction—known as the force of gravity—depends only on mass. Enormous amounts of mass let the gravity of galaxies hang on to stars, stars hang on to planets, and planets hang on to their moons. Earth's force of gravity (referred to as 1-*g*) is strong enough to keep us and everything else on the surface held in place. Spaceships that are launched into orbit are in a state of constant free fall, and occupants inside experience weightlessness or zero-*g* as long as the ship is not accelerating or decelerating (which is why spinning ships can use centrifugal force to simulate 1-*g*). Travelers to the Moon are visiting a world with much less mass than Earth, and thus the force of gravity is much less (specifically, the Moon's low-*g* is only about one-sixth of Earth's gravity). Mars has a stronger gravitational force than the Moon but is still low-*g*, at about three-eighths that of Earth. Tiny worlds like Phobos, Deimos, and the Near-Earth and Main Belt Asteroids have extremely low-*g* or even "microgravity" levels, ranging from only about 0.01 to 3 percent of Earth's gravity. Hang on carefully in those environments, lest you accidentally jump off into space!

you're taking a 1-*g* rotating ship to your destination. If you are booking a luxury trip with formal-dress opportunities, check with your spaceline, as they may have tuxedo and gown rental options.

Electronics While many professional photographers will carry specialized equipment, in general you won't need more than a standard handheld or space-suit camera for photographing the spectacular sights of the universe. Alternatively, if you have bioelectronic vision enhancements, you will likely be within interplanetary standard wireless-signal ranges to store and stream your photographic and video memories for all but the most distant solar system destinations.

Medicine Be sure to remember any specific medicines unique to your health situation, as well as over-the-counter medical supplies to combat standard space-related illnesses like dizziness or nausea from changes in gravity. Most of the major spacelines and resort destinations have well-stocked pharmacies and tourist medical facilities, but you'll pay hefty prices for the convenience.

But how about preparing for off-ship destinations? Here the advice is simple: Unless you're an experienced traveler with your own specialized space suit and life-support equipment, rely on the locals! Conditions vary widely across the solar system, including extremes of temperature, pressure, and radiation. Local tour experts have spent many decades perfecting exactly the right space suits, individual travel pods, and small-group tour vessels for each of these destinations. We highlight some of this special insider information in the "Local Flair" section of each chapter. Put your trust in their skills and experience.

TRAVEL AND ACCOMMODATIONS

There are dozens of licensed commercial spaceline transport vehicles on which you can book accommodations for the outbound and (unless you're planning a one-way trip) return legs of your adventure. Rooms on these vessels range from economy to luxury—from tiny hibernation pods where you can sleep for most of your trip, to multiroom, king-size suites with extra-large view ports. The sky is (literally) the limit, and the choices really depend most on how much you're willing to spend and how long you want to spend traveling. Cruise times to some solar system destinations can be relatively short (days to weeks), while others can be quite long (several years or longer). Traveling beyond the solar system to the nearest stars will take even longer—for example, the first planned interstellar trip to the TRAPPIST-1 system, 40 light-years from Earth, could take 80 years or more! So, be sure to factor your patience and your budget—as well as your life expectancy—into your planning for the cruise segment of your trip.

You may also wish to consider one of the many options for extending your stay in Earth's orbit prior to departing for deep-space destinations. Here, again, the options range from spartan to opulent. Long gone are the early days of cramped accommodations in school bus–size orbiting tin cans. Hundreds of years of experience and experimentation by the hospitality industry have resulted in orbital hotel and resort options that have been optimized for singles, couples, and families of all ages and backgrounds. Whether you're looking for a few days of quiet reflection while enjoying the low-orbit view of our glorious blue planet passing below, a romantic stargazing escape for a week in a high-orbit geostationary honeymoon resort, or a multiweek family adventure at a zero g theme park, there are plenty of choices. And if the Moon is not your ultimate destination, consider

EARTH Your Oasis in Space
WHERE THE AIR IS FREE and BREATHING IS EASY

Traveling off-world will surely give you a much greater appreciation for our unique home planet.

ASTRONOMICAL UNITS

The scale of the solar system is so much larger than the distances we typically deal with on Earth, that it can be hard to grasp. We can relate to distances of hundreds or even thousands of miles, but millions? Or billions? To try to make those distances more imaginable and intuitive, astronomers invented the concept of the astronomical unit, or AU. One AU is the average distance between the Earth and the Sun, or about 93 million miles (150 million kilometers). Mercury orbits close to the Sun at an average distance of about 36 million miles (58 million kilometers), equal to about 0.39 AU. Mars is farther from the Sun, orbiting at about 1.5 AU. Bright comets on highly elliptical orbits that take thousands to tens of thousands of years to orbit can travel out many hundreds to many thousands of AU before slowly falling back in toward the Sun. And the gravitational influence of the Sun itself is thought to extend out to around 30,000 AU, a distance roughly equal to a light-year (the distance traveled in 1 year at the speed of light, which is 186,000 miles per second, or 300,000 kilometers per second).

an extended stay at one of the lunar bases as a pre-trip add-on (see chapter 1 for more details).

Once you reach your destination, common ways to get around include continued orbital touring in your spaceliner, short-segment trips to and from surface or atmospheric destinations on shuttles designed specifically for each environment, surface-based rovers or submarine vessels (on Europa), and of course just plain old hikes in a properly outfitted space suit. To assist you on your adventures, we've included maps of some of the most popular attractions and landmarks, with gold stars indicating bases, research centers, tours, and recreational opportunities. Special "Don't Miss . . ." sections in each chapter guide you through the most unique and enjoyable tourist activities, with easy-to-identify icons indicating what is available at each one.

When considering local accommodations, be sure to plan ahead. Certain options may only be available with significant advance planning. Room is often limited, for example, on special excursions to destinations with extreme environmental constraints (see section on Local Traditions and Conditions below) or which require specific timing or placement for events such as eclipses or geyser eruptions. In this book, we outline as many of these choices as we have been able to identify. Some destinations, however—especially those at out-of-the-way moons or asteroids that have only recently become popular as tourist attractions—may still have only one accommodation choice, and it could be quite basic! Do your homework in advance.

LOCAL TRADITIONS AND CONDITIONS

Finally, remember that you are leaving your home planet—the cocoon that you were literally made to live in—and subjecting yourself to extremely harsh alien environments that can literally kill you if you're not prepared. Take the time to learn about the hardships to which you, your fellow crewmates, your support staff, and your equipment will be subjected. Participate fully and attentively in your pre-trip safety and equipment training. Get in the best physical shape you can prior to

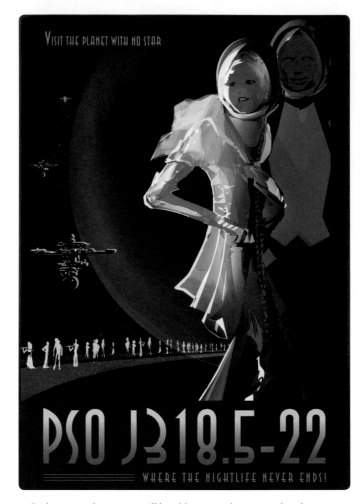

Perhaps one day soon you'll be able to travel to extrasolar planets such as PSO J318.5-22, a rogue planet without a host star.

the trip, so you will be ready for what could potentially be the most physically demanding experience of your life. Understanding what you and your equipment are capable of could save your life or the lives of others in the event of unanticipated troubles.

Traveling in space is *never* routine, and neither should be your preparation for the experience. As we describe the opportunities for exploration and adventure at each of our solar system's many exciting destinations, we'll also use the latest scientific information available to give you a good feel for the atmospheric, surface, and/or subsurface conditions that you're likely to encounter, either during your transport to and from those destinations or during your time "out in the wild," so to speak. Pay attention to the world around you, stay safe and diligent, and hang on for the ride!

KEY TO ICONS USED IN THIS BOOK

In the "Before You Go" and "Don't Miss . . ." sections of this book, you'll find icons to help you identify dining and recreational opportunities, as well as potential pitfalls to be aware of. These include:

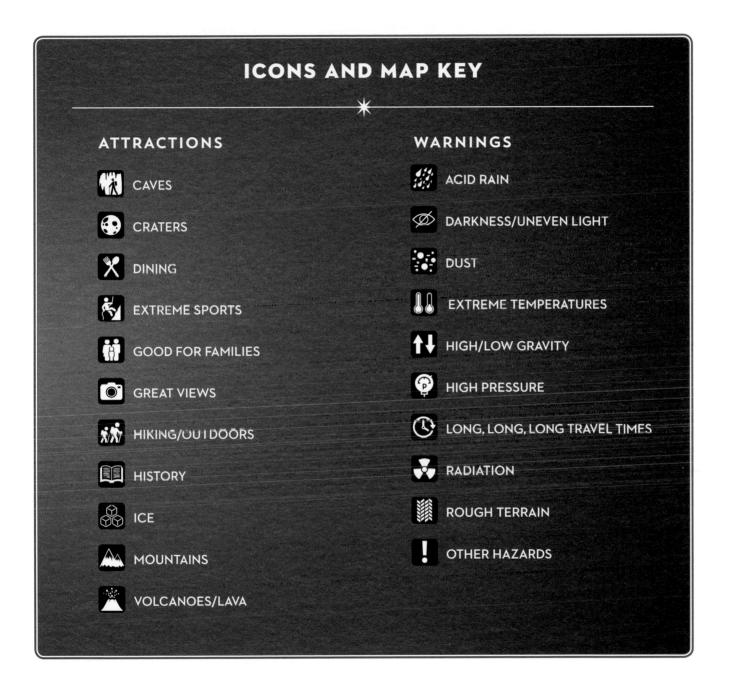

ICONS AND MAP KEY

ATTRACTIONS

- CAVES
- CRATERS
- DINING
- EXTREME SPORTS
- GOOD FOR FAMILIES
- GREAT VIEWS
- HIKING/OUTDOORS
- HISTORY
- ICE
- MOUNTAINS
- VOLCANOES/LAVA

WARNINGS

- ACID RAIN
- DARKNESS/UNEVEN LIGHT
- DUST
- EXTREME TEMPERATURES
- HIGH/LOW GRAVITY
- HIGH PRESSURE
- LONG, LONG, LONG TRAVEL TIMES
- RADIATION
- ROUGH TERRAIN
- OTHER HAZARDS

1

WEEKENDING ON THE MOON

These of us living on Earth are incredibly lucky to have such a large and interesting space-travel destination so close to home. The Moon is Earth's closest celestial companion, and makes the Earth the only inner-solar-system world with its own planet-size natural satellite (while the Martian moons, Phobos and Deimos, are fun places to visit—see chapters 5 and 6—they are really tiny compared to Mars itself). The Moon is about a quarter the diameter of Earth, and its total surface area is a little less than the size of Asia. That makes for a lot of places to discover and explore!

Opposite: Our nearest celestial neighbor is no longer a destination just for lucky astronauts or miners. So pack up, blast off, and plan to spend a weekend or more exploring the Moon! (*Moon*, illustration by Indelible Ink Workshop)

People have been dreaming about visiting the Moon for thousands of years. Those dreams became a reality—at least for 12 people—in the late 1960s and early 1970s when NASA's Apollo program sent the first astronauts to visit and study our nearest neighbor. They brought back rock and soil samples and other data that helped us to learn that the Moon was formed only 30 to 50 million years after the Earth formed, and probably by a cataclysmic event known as a giant impact. Imagine what the scene must have been like, more than 4.5 billion years ago, when a Mars-size planet crashed at a grazing angle into the young Earth, scattering rocky and melted debris in all directions. As the Earth cooled and recovered from this shock, some of that scattered debris went into orbit and eventually coalesced to form our Moon. What a wild ride!

The old scars of that ancient impact—as well as fresh scars made by millions more asteroids and comets crashing onto the Moon's surface—are among the many different kinds of exotic geologic features waiting to be explored on the Moon. And while you still need to have a good space suit and a supply of your own oxygen like our twentieth-century ancestors did to survive out on the airless lunar surface, you don't need to be a trained astronaut to set out on your own lunar adventure!

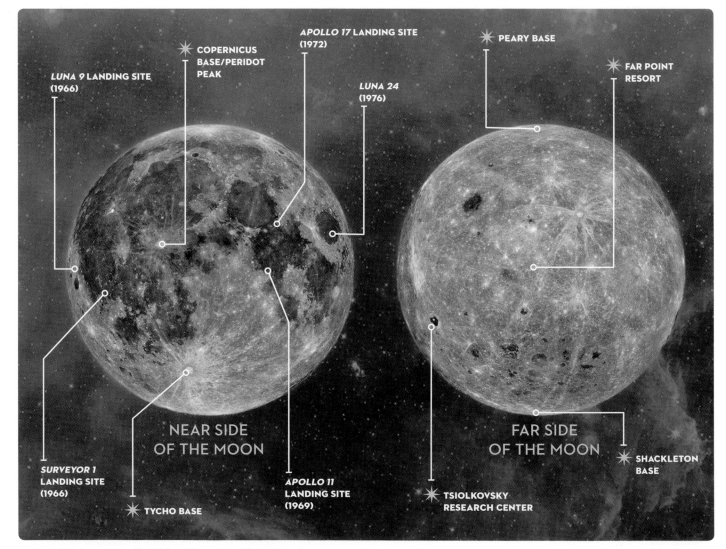

Map of the lunar near-side (left) and far-side (right), showing the locations of the major lunar bases as well as some of the attractions described here. Starred items represent bases and attractions.

MOON FAST FACTS

Type of Body
Planetary satellite (moon)

✳

Distance from Sun
Average is ~1 AU, or 93 million mi.
(150 million km)

✳

Distance from Earth
Averages 240,000 mi. (384,000 km)

Travel Time from Earth
Varies from ~6-8 hours to ~3 days

✳

Diameter
2,159 mi. (3,474 km), about 27% the
diameter of Earth

✳

Highlight
Humanity's first stepping-stone into
the cosmos—a must-visit!

AVERAGE TEMPERATURES

	DAYTIME HIGH		NIGHTTIME LOW	
	°F	°C	°F	°C
Mean Surface	225	107	-243	-153
Equator (°0 Latitude)	252	122	-252	-158
Midlatitudes	171	77	-225	-143
Poles	-45	-43	-81	-631
Dark Polar Crater	-387	-233	-387	-233

BEFORE YOU GO

Despite the fact that thousands of people now live and work on the Moon, it is still a harsh and rugged environment in which—just like many other forms of adventure travel or extreme-sports activities—it is easy to get hurt or even killed if you're not properly prepared. For your various lunar explorations, you'll want to be prepared for the following:

🌡 **Extreme Temperatures:** Temperatures on the Moon range anywhere from -387 to 252 degrees Fahrenheit (-233 to 122 degrees Celsius), depending on the time and location. To protect yourself when you're outside a thermally controlled habitat or base, be sure that your space suit has been designed for this environment. Unless you're an experienced traveler with your own specialized suit and life-support equipment, the best way to make sure your suit is up to the challenge is to rely on the locals! Lunar engineers and tour experts have spent centuries perfecting exactly the right space suits, individual travel pods, and small-group tour vessels for your time on the Moon, so you should feel at ease putting your trust in their skills and experience. Be sure to check on your options ahead of time and reserve the appropriate equipment lest you be left in the ship.

⬆ **Low Gravity:** Because the Moon is a smaller world than the Earth, it has lower surface gravity (g); on the Moon you will feel like you weigh only about 16 percent of what you weigh back home! While zero-g and low-g destinations might seem like great weight-loss solutions, there are many challenges to deal with as you work to get your "space legs."

To prepare for the low-g conditions you will encounter on the way to and on the Moon, pack light and focus on the essentials: light and informal clothes for comfortable zero-g or low-g cruising, and good gripping socks (many still prefer the old Velcro variety) for standard low-g tunnels and corridors in some of the larger vessels.

If you experience space sickness on the trip from Earth (as many people do in zero-g), be sure to set aside time for rest and recovery. Factor in at least two days at the beginning of your trip to acclimate to the lunar environment, while spending some quality time staring at that spectacular blue marble always hovering in the inky black sky.

Rough Terrain: Craters and other uneven surfaces—combined with the low gravity—can lead to cuts and scrapes from falls for "gnewbies," as you will be lovingly called. Bring plenty of elbow, wrist, and knee pads for protection, and make sure your space suit is outfitted with extra padding as well, especially on your first few excursions.

Darkness: The moon is in darkness for half of the lunar day, which is about 350 hours out of the Moon's approximately 700-hour orbit around the Earth. There are a variety of ways to prepare for spending time in the dark on the Moon. Like nocturnal animals back on Earth, many people choose to burrow underground and partake in the many indoor activities available in the lunar bases, which maintain 12-hour day/night schedules controlled through artificial lighting. Thermal control keeps the temperatures stable in the bases whether it's day or night outside. Or you can brave the extremely cold nighttime temperatures and take a night tour of Apollo Park, Luna Park, or any of the major hiking trails. The crowds will be smaller during the lunar night, but the requirements on your space suit and other equipment will be stricter because of the extreme cold.

DON'T MISS . . .

Whether you're visiting for just the weekend, planning a week or two stay, or even thinking about retiring to the lunar surface, there are some incredible sights and experiences that you simply MUST try to get onto your itinerary. These include:

THE APOLLO INTERNATIONAL HISTORIC PARK

Start your lunar tour with the six places on the Earth-facing side of the Moon visited by the first astronauts. Whether you choose to visit just one of the Apollo landing sites (most visitors choose the first, the landing site of the *Apollo 11* mission's *Eagle* lander in the Sea of Tranquility, made famous by astronauts Neil Armstrong and Buzz Aldrin), or to tour them all in historically accurate order, park rangers at each site's visitor center will fill you in on the historical context and major scientific achievements of these culturally important sites.

The sites have been painstakingly restored to their original post-Apollo states, and they are now protected by a transparent dome field to help prevent additional erosion and vandalism. Landing modules, rovers, flags, and science experiments are arrayed just as the astronauts left them in 1969 through 1972. While you can't walk around among the artifacts themselves, you can sign up for a special tour of the inside of the dome, using new

Buzz Aldrin's boot print in the fine, powdery lunar soil.

hover-grav technology that does not interact with the surface. Ranger pilots will take you right up to the landing craft, the famous lunar rovers, and along the longer traverses taken by the later missions. Be sure to sign up early, however, as tours routinely sell out.

The landing sites of the five successful NASA Surveyor robotic landers (which landed on the Moon between 1966 and 1968) were also recently added to Apollo Park, making this site a must for closet historians and science buffs!

LUNA PARK

If you can't get enough of early-space-program lunar robotic artifacts, also check out Luna Interplanetary Historic Park. So far, the park includes five historic sites where, between 1970 and 1976, the USSR's *Luna 16*, *20*, and *24* landers gathered samples from the Moon robotically, and the *Luna 17*'s *Lunokhod 1* rover and *Luna 21*'s *Lunokhod 2* rover explored the surface. The Luna Park sites can be visited as quick stops, but park rangers offer fun excursions that follow the lengthy traverse paths of the Lunokhod rovers and let you re-create the obstacle avoidance moves that their near-real-time controllers back on Earth had to make.

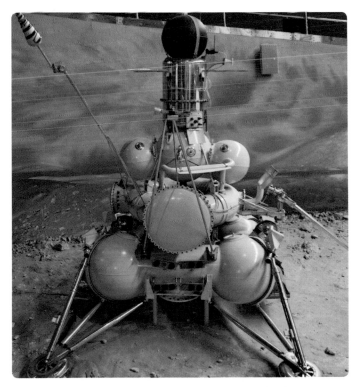

A model of a Soviet *Luna* sample-return lander, used for three robotic lunar sample-return missions in the 1970s.

THE FIRST LUNAR LANDING SITES

The first aliens on the Moon weren't astronauts from planet Earth, but instead were robotic emissaries sent from the United States and the former Union of Soviet Socialist Republics (USSR) to study the lunar surface and scout for potential landing sites for future astronauts. Specifically, between May 1966 and January 1968, the US sent seven robotic Surveyor landers to the Moon. Five landed successfully, and one (*Surveyor 3*) ended up being the landing site for the *Apollo 12* astronauts.

And between January 1963 and August 1976, the USSR launched a whopping twenty-seven robotic missions intended to land on the Moon. Of these, only eight were successful (the first in January 1966), but they included the first robotic missions (three of them) to return samples of the Moon to the Earth, and the first robotic rovers (two of them, in 1970 and 1973) to drive around on another world. These early reconnaissance sites are slowly being added as satellite nodes of the Apollo and Luna Park systems.

POLAR ICE MINES AND NEARBY ATTRACTIONS

Huge deposits of water ice and other ices are buried near the north and south poles of the Moon. Mining, processing, and distribution of those natural resources to the colonies and other facilities on the Moon (and elsewhere in the solar system) is a major lunar industry, which makes visiting the poles a major tourist destination.

Ice Mines

The several available mine tours are run by professional agencies with (so far) spotless safety records. People who've taken the tours almost uniformly rave that the experience is well worth the out-of-the-way trip to the far north or south of the Moon. Enormous elevator cars (initially used for hauling ice from veins that are now mostly spent) have been retrofitted for comfort and take tourists down hundreds of meters below the surface, into a sealed environment of caves and passages that have been filled with normal atmosphere and heated to shirt-sleeve temperatures. Tour operators show off some of the original, late twenty-first-century equipment used to dig out and transport the ice, as well as live feeds showing modern equipment being used in the tourist-inaccessible parts of the mines. If you're staying for a while, you can learn how to low-g ice skate on some of the many rinks that are popping up closer to the residential parts of the mine communities.

Dining

Some spectacular bars, pubs, and restaurants have cropped up recently within the underground mines. Look especially for dining and drinking establishments that offer the popular (albeit expensive) brand-name Lunar

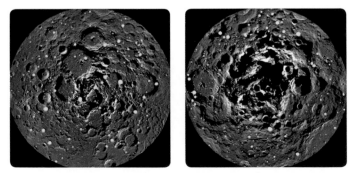

Two views of the lunar polar regions showing the surface from the north pole (left) and south pole (right). Yellow dots show regions in permanent shadow where ice has been suspected or discovered.

ICE ON THE MOON?

The Moon, like the Earth, is relatively close to the Sun compared to most of the rest of the solar system. Both planets dwell in the warm zone where liquid water can be stable under the right conditions (like those on Earth). But unlike the Earth, the Moon does not have an atmosphere to help keep that water warm and in the liquid state. Without air, the surface of the Moon is like the inside of a vacuum jar, and any liquid water on the surface quickly evaporates into water vapor. Even water ice can't survive exposed to sunlight on the Moon's surface, as the daytime temperatures are warm enough to evaporate away any ice on the surface.

It may come as a surprise, then, to learn that there is ice buried in the shallow subsurface near the Moon's north and south poles. Why doesn't it evaporate away in the sunlight? Well, the key is that there are places near the poles of the Moon where, literally, the Sun never shines. If the Moon was perfectly flat, then the Sun would always be just slightly at or above the horizon at the north and south poles, similar to the way the Sun is always low in the sky in Alaska or Antarctica on Earth. However, the Moon is not flat. Impact craters near the poles are deep holes in the ground surrounded by raised mountainous ridges—the craters' rims. For many of these craters, the Sun never gets above the local crater rim horizon, and so the bottoms are permanently shadowed. Even though the Moon is relatively close to the Sun, where it's dark, temperatures are extremely low (hundreds of degrees below 0 degrees Fahrenheit), and so any ice that's there won't evaporate.

But how did the ice get there? Planets are constantly being bombarded by rocky asteroids and icy comets, which are the small leftovers from the formation of the solar system 4.56 billion years ago. When a small icy comet crashes into the Moon near the equator, sunlight will eventually evaporate any remaining icy debris into space. But if one happens to crash into one of the permanently shadowed craters near a pole, any remaining ice will stay there in the deep cold, for a long, long time.

These icy polar deposits were discovered in the late twentieth century by some early NASA robotic space probes. Scientists have been studying the ice ever since to learn more about the history of the solar system (in addition to water ice, the comets also bring other ices such as CO_2 (dry) ice, methane ice, and nitrogen ice). These regions have also become important natural resources for lunar explorers and others. Indeed, ice is now the "gold" of the solar system, because of its value to supply water, radiation shielding, hydrogen and oxygen for rocket fuel, and oxygen for breathing.

Origine Contrôlée (LOC) ice and water as integral parts of their food and beverages. LOC water and ice come from the most pristine (with little or no contamination from surrounding rock and soil) subsurface deposits. Beverage aficionados and food critics on the Moon and Earth (a major export destination for LOC ice and water) are going bonkers about not only what they claim is the superior taste of these cosmic deposits, but also their marketing potential. From where else can you offer customers the ability to drink 4.5-billion-year-old water or a whiskey on the rocks—where these particular "rocks" are older than the oldest rocks exposed on Earth?

PERIDOT PEAK HIKE AT COPERNICUS CRATER

For the more active interplanetary traveler, the Lunar Trekking Network has organized an enormous number of hikes, from easy to strenuous, on the Moon. Some of the most spectacular are the climbs up any of the central peaks of the large impact crater Copernicus, near the center of the near side of the Moon. Back in the early days of the space program, astronomers discovered that the mountains here are enriched in the mineral olivine, an iron-bearing silicate that forms deep in the mantle of rocky planets like the Moon (and the Earth). The asteroid impact that created Copernicus crater (estimated to have occurred about 800 million years ago) caused an explosion that created the 57.7-mile-wide (93-kilometer) crater and excavated some of that olivine-bearing mantle material. Central peaks of large impact craters represent the materials that were deepest underground before the crater formed, which is why some (like at Copernicus) contain olivine. Gem-quality olivine on Earth is called peridot, and while gem-quality olivine is very rare on the Moon, hikers have nonetheless given the trail up the tallest of Copernicus crater's central peaks a fanciful association with that gemstone.

The Peridot Peak Trail winds up the 0.75-mile-tall (1.2-kilometer) mountain via many switchbacks, which help to keep the slopes reasonable as well as to bring in numerous panoramic vistas of the floor and walls of

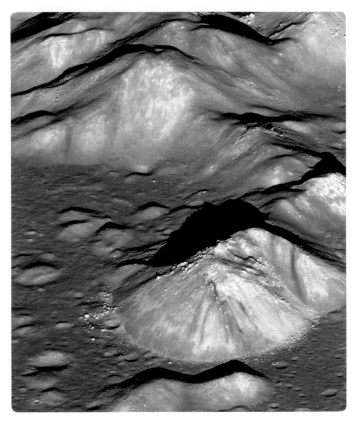

A view of Copernicus crater's central peaks. The 2.5-mile-wide (4-kilometer), 2,300-foot-tall (700-meter) mountain near the middle of this view has been informally named Peridot Peak by the local hiking club.

PROTECTING LUNAR LANDING SITES

Back in the late twentieth and early twenty-first centuries, a bevy of new, smaller, commercially led robotic missions were sent to explore the Moon. Some of these began landing and roving on the Moon, and a few trundled into special places like the original *Apollo* landing sites. There was significant public and governmental outcry about the potential—even for well-meaning explorers—to desecrate important human cultural sites and artifacts.

Even though international treaties such as the famous Outer Space Treaty of 1967 stipulate that the artifacts (landers, rovers, flags) brought to the Moon remain the property of the nations that sent them there, that and other treaties since do not allow any person or nation to claim rights to ownership of parts of the Moon or other celestial bodies. So, how could they be legally protected, and by whom?

After significant multilateral efforts and negotiations, in the mid-twenty-first century, the United Nations agreed to put the *Apollo* and *Luna* landing sites under the protection of the same kinds of international protocols that protect places like Antarctica (which is also not part of any specific nation), and the landing sites were declared UNESCO (United Nations Educational, Scientific and Cultural Organization) Interplanetary Scientific Heritage Sites by unanimous international consent.

the crater and the distant plains beyond. Even if the Sun is relatively low in the sky and much of the trail is in shadow, earthshine is usually bright enough to light your way without the need for a headlamp. Photographers will especially enjoy the spectacular vistas. And who knows, perhaps you'll kick up a peridot gemstone along the way . . .

LAVA TUBES OF ARISTARCHUS PLATEAU

If you've visited the Hadley Rille (*Apollo 15*) section of Apollo Park, you've seen a lava tube: a long, sinuous conduit where a river of molten rock once flowed across the surface. Hadley Rille is an open lava-tube channel now, as the original top covering of the tube has long since eroded away. There are other places on the Moon, however, where you can explore intact lava-tube caves that can take you deep underground into a lunar geologic wonderland. Perhaps the most famous is the lava-tube network in the Aristarchus Plateau.

The Plateau is a bright, elevated region of concentrated, ancient lunar volcanic activity set against the dark Ocean of Storms and near the near-side crater Aristarchus. Several large eroded lava tubes (the largest can easily be seen with telescopes on Earth) wind sinuously through this area, and early lunar-surface geologic expeditions quickly discovered a vast interconnected network of subterranean lava tubes that carried molten rock down into the surrounding plains. When the eruptions stopped (billions of years ago), the lava drained out and cooled, leaving behind spectacular tunnels to explore. Several reputable tourist companies now offer expeditions into that lava-tube network, where you can see and learn about the complex and still somewhat mysterious geologic history of this part of the Moon.

Apollo 15 astronaut Dave Scott parked the lunar rover on the rim of Hadley Rille back in 1971. Time limits for their excursions prevented the astronauts from exploring the canyon below.

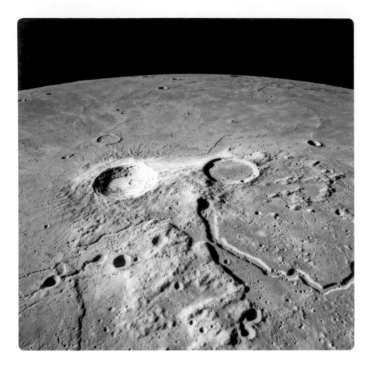

Shopping and Entertainment

A few years ago, a consortium of adventure tourism companies obtained permission to go several miles into one of the largest lava tubes and seal off the rest of the tube to create a warm, breathable environment, now called TubeLand. They created a low-g theme park and a mall-like area for shopping and dining near the entrance. Families with young children or teenagers will find lots of fun activities there. Deeper into the tube, the consortium created special areas for camping and hiking, and there's even a special low-g network of biking trails for the truly adventurous.

The winding channel seen here, known as Schröter's Valley, is an open-roof lava tube that ranges from about 3 to 6 miles (5 to 10 kilometers) wide.

HISTORY OF EXPLORING THE MOON

1839: First photographs of the Moon (daguerreotype plates)

1959 (Sept.): First human object to reach the Moon (USSR *Luna 2* impactor mission)

1959 (Oct.): First photographs of the far side of the Moon (USSR *Luna 3* flyby mission)

1966: First soft landing on the Moon (USSR *Luna 9* robotic lander)

1969: First human landing on the Moon (US *Apollo 11* mission)

1970: First robotic sample-return mission to/from the Moon (USSR *Luna 16* mission)

1972: Last (of six) Apollo-mission human landing on the Moon (US Apollo 17 mission)

2009: First confirmation of ice deposits at the lunar south pole (US *LCROSS* mission)

2030: First *Lunar Scout* human landing on the Moon (joint US, Russia, China)

2033: Early human and robotic landing sites declared UNESCO Interplanetary Scientific Heritage Sites

2035: First private "tourists" orbit and visit the Moon

2050: First lunar orbital science/research station established (US, Japan, others; *Selene* station)

2071: First lunar colony established, at Shackleton crater (near south pole)

2080: First commercial ice mining opens near Shackleton base

2095: First routine spaceline flights begin to ferry tourists to the Moon

2176: Number of annual tourists visiting the Moon exceeds 10,000

2218: Plans announced for first major far-side lunar base, in Tsiolkovsky crater

DAY AND NIGHT ON THE MOON

The Moon orbits Earth in what astronomers call *synchronous rotation*. That is, the Moon spins on its axis exactly once for every orbit it makes around the Earth (this orbit takes 27.3 Earth days). From our perspective on Earth, it doesn't look like the Moon is spinning at all, though, because this synchronous rotation means the same hemisphere (or "face") of the Moon is always pointed toward Earth. That's why our familiar view of the full Moon is what astronomers call the *near side*, the side always facing toward us; the side always facing away is the *far side*.

From your perspective on the Moon, you will notice that Earth is about four times larger than the full Moon is as viewed from your home back on Earth. If you're near the Moon's equator, Earth will be high in the sky, and if you're near the poles, Earth will be low and close to the horizon. But, unlike the Moon as seen from Earth, Earth as seen from the Moon doesn't move across the sky -it stays there, in the same parts of the sky all the time, because of synchronous rotation.

Despite how it looks, the Earth *does* change: It goes through phases, just like the Moon does

as seen from Earth. When it is full Moon on Earth (Sun, Earth, and Moon are aligned, in that order), it is "new Earth" from the Moon. And vice versa: new Moon from Earth is "full Earth" from the Moon. And in between, the new Earth becomes a thin crescent, then a quarter Earth, and then a full Earth—before shrinking to third-quarter Earth, waning-crescent Earth, and back to new Earth again, every month.

The Moon changes, too. Because it is spinning on its axis, every spot on the Moon goes through day and night just like every spot on Earth, except a "day" on the Moon lasts for about 13.5 Earth days of daylight and about 13.5 Earth days of darkness. Most tourist activities (just like the early Apollo missions) are timed to occur during the lunar day when the Sun is up and temperatures are warmer. Much of the activity then transitions to the subsurface for a few weeks while the Sun is set. Interestingly, there's even a subculture of lunar inhabitants known as "sunbirds" who constantly follow the Sun, switching between near-side and far-side homes on a biweekly schedule.

Below: Earth as seen from the Moon.

FAR SIDE OF THE MOON

Tired of the Earth? Never want to see it again? Then the far side of the Moon is the perfect place to get away from it all. The isolation from Earth, relatively undeveloped environment, and stunning views of the Earth-less night sky during the long lunar nights (which last 13.5 Earth days; see box on p. 11) provide opportunity for quiet, out-of-the way, and often deeply transcendent experiences.

Far-side accommodation options are much fewer in number than on the near side, and they are dominated by private homes, scattered campgrounds, and a few resorts. Many far-side destinations offer artistic-themed experiences, including getaways where you can focus on painting, music, dance, astrophotography, local geologic exploration, and even cooking classes. At the Interplanetary Scientific Foundation's Tsiolkovsky Research Center, for example (in the Tsiolkovsky crater, just a short hop onto the far side and soon-to-be the newest major lunar base), you can sign up for radio astronomy research tours and nighttime astrophotography lessons under the darkest, clearest skies you might ever witness.

Getting to your ultimate destination might require patience (for example, booking passage on a weekly cargo ship scheduled to resupply your destination might be the only way to get there and back). Still, the vibe is quite different from the near side, and if getting off the beaten path is your goal, the far side of the Moon is still largely terra incognita.

GETTING THERE

Getting to the Moon should not be a problem as long as you book in advance. Options range from economy to first class, depending on your budget, and travel times vary widely. On the slow end, you can reach the Moon using traditional chemical propulsion technology quite similar to that used on the Saturn V rockets that brought the first astronauts to the Moon back in the twentieth century. Just like the Apollo astronauts, it will take you about three days to get out to the Moon using traditional rocket technology,

and then another three days to return to Earth once your time on the Moon is over. Using newer technology such as nuclear propulsion engines or hybrid nuclear-chemical rocket technologies, you can cut down on the transit time significantly. The fastest shuttles can get passengers to or from the Moon in as few as six hours, although the ticket prices are substantially higher than for the slower trips.

It's sometimes challenging to get around on the Moon itself, depending on how remote your destination is. Most of the major tour companies offer charter service to the most popular tourist destinations, oftentimes using small shuttles built specifically for the lunar gravity and environment. For example, the new LunaTransit Corporation's S-2200 shuttles are built on the Moon from light materials that only have to withstand lunar gravity, not Earth's stronger pull. Less weight allows these shuttles to lift off and maneuver using much less fuel than terrestrial shuttles, while still providing comfortable accommodations and meeting the rigorous thermal, atmospheric, and radiation safety standards expected for lunar transport vehicles. Plans are in the works for a local government-run lunar transportation network, but implementation is still many years away.

THINGS TO DO, PLACES TO STAY

Attractions and accommodations abound for families, couples, and individual travelers across a wide range of comfort levels and prices. The list of hotels and resorts on the Moon is too long to list, but you can generally assume that the standard terrestrial chains are all represented at the various bases and settlements. Still, there are unique and eclectic choices. For example, Endless Sunset Resort, a sprawling complex on the permanently sunlit part of the rim of Shackleton crater, offers amazing panoramic views from every room. For the more budget-minded, Camp Moon Inc. has just opened a new set of domed trek huts near the Hadley Rille trailhead in Mare Imbrium. If you've ever wanted to take that lava-tube hike that astronauts Dave Scott and Jim Irwin never got to do back in the summer of 1971, here's your chance!

If fine dining is your thing, you'll find a range of establishments and cuisines as diverse as those on Earth

LUNAR SOIL

Like the land areas of Earth, much of the Moon's surface is covered in a fine layer of rocky debris that planetary scientists call the lunar soil. Lunar soil is created by the physical and chemical weathering of lunar rocks, just like soil on Earth; however, impact craters play a very important role in soil formation on the Moon. While Earth's soil is strongly influenced by life—microbes and plants promote soil formation, break soil down into finer fragments, and inject nutrients into the soil—the Moon has always been lifeless, and so there is no biologic component to the lunar soil. That said, many of the parent rocks on the Moon are similar to the parent rocks of many of Earth's soils. Specifically, both the Earth and the Moon (as well as Mercury, Venus, Mars, and some of the larger asteroids) have similar basaltic (high iron and magnesium, and low silicon) volcanic rocks on their surfaces. These volcanic rocks can be broken down to form soils. Thus, when biologic components—nutrients from plants, microbes, and other organic matter—are added to the lifeless lunar soil, the result is a mixture that is a lot like many of Earth's soils. While it took planetary botanists many decades to perfect the right recipe, plants grown in properly cultivated lunar soils can routinely produce outstanding yields of all kinds of fruits and vegetables. Because of small trace mineral differences, however, discerning foodies can quickly learn to recognize and even savor the subtle differences in taste between lunar-grown and Earth-grown produce.

waiting for you on the Moon. However, be prepared for some subtle but (usually) quite enjoyable shifts in tastes and flavors due to the differences in gravity, temperature, pressure, soil composition, and other factors. Lunar cuisine has emerged as a unique and exciting contrast and complement to its Earth-bound relatives.

Active travelers who want to get out and about will find numerous opportunities on the Moon. Hiking, camping, underground-cave biking, jet-pack trekking, ice skating, and even low-*g* sports complexes have emerged as fun and exciting lunar tourist activities. Join the Shepard Golf Club, named after *Apollo 14*'s commander, Alan Shepard, who hit the first golf balls on the Moon back in 1971. Or camp under the most spectacular Milky Way sky you've ever experienced by staying at any number of remote far-side getaway spots.

Whether it's a weekend or a lifetime, you'll never run out of exciting places to see and experience on our nearest celestial neighbor!

LOCAL FLAIR

The families of some "Loonies" (as some people call themselves) go back nearly 150 years to the first lunar colonists of the mid-twenty-first century. You might notice that many of these folks are extremely tall and thin, and they seem to move with a unique grace and flow in the lower lunar gravity. Many are *truly* locals, born on the Moon and part of a new genetic pool of humans who are slowly adapting and, in very minor ways, even evolving to the specific conditions (especially the lower gravity) of the lunar environment. Early evolutionary biologists realized that humans would adapt and evolve once we became a multi-planet species, but it was impossible to predict exactly how those adaptations would manifest. Well, this experiment in human evolution is now running on the Moon (and on Mars and other off-world destinations), and it is fascinating to meet and learn from the locals all the different ways that their lives, and their bodies, now differ from those of their terrestrial ancestors. Take advantage of this unique cultural opportunity!

HEATING UP ON VENUS

Long known as Earth's "sister planet" prior to the space age, Venus has turned out to be a fraternal rather than identical twin to our home world. There are many similarities, of course, between Earth and our nearest planetary neighbor: similar size and gravity, similar overall rocky and metallic composition, similar position in the inner solar system—only about 25 percent closer to the Sun. But the similarities end there.

While both worlds have atmospheres, they are dramatically different ones. The skies on Venus are filled with immense quantities of carbon dioxide (CO_2), and the surface pressure is more than 90 times the Earth's surface pressure—equivalent to the pressure felt in a submarine 3,000 feet (900 meters) underwater. That thick atmosphere sports violent upper-level wind speeds of more than 220 miles per hour (350 kilometers per hour).

Opposite: Floating airhotels have become all the rage in the temperate, almost Earthlike upper levels of Venus's atmosphere. Book your stay today!

The result of all that CO_2 is global warming gone wild. As your transport descends to the surface, the temperatures rise to more than 870 degrees Fahrenheit (460 degrees Celsius), much hotter than your oven and hot enough to melt lead! Another piece of individuality comes from the way Venus spins on its axis: Venus is barely spinning, taking about 243 Earth days just to spin once on its axis. And, for reasons no one understands, it's spinning backward relative to the Earth and most of the other planets! Strange world indeed.

But don't let those oddities and extremes stop you from visiting the solar system's hottest hot spot outside the Sun. A visit to Venus will give you an experience unlike anywhere else in the solar system, whether you choose to float lazily in one of several balloon-borne airhotels located in the Earthlike temperatures of the upper atmosphere, to go adventure skydiving through clouds of sulfuric acid, or to don a hotsuit and brave the hellish conditions of the planet's volcanically active surface, including the rugged mountains of Ishtar or Aphrodite Terra.

BEFORE YOU GO

The technologies needed for safe touring on and above Venus only became widely available in the late 2100s, and tour companies are still experimenting with new designs for accommodations and tourist excursions. For your various Venus explorations, you'll want to be prepared for the following:

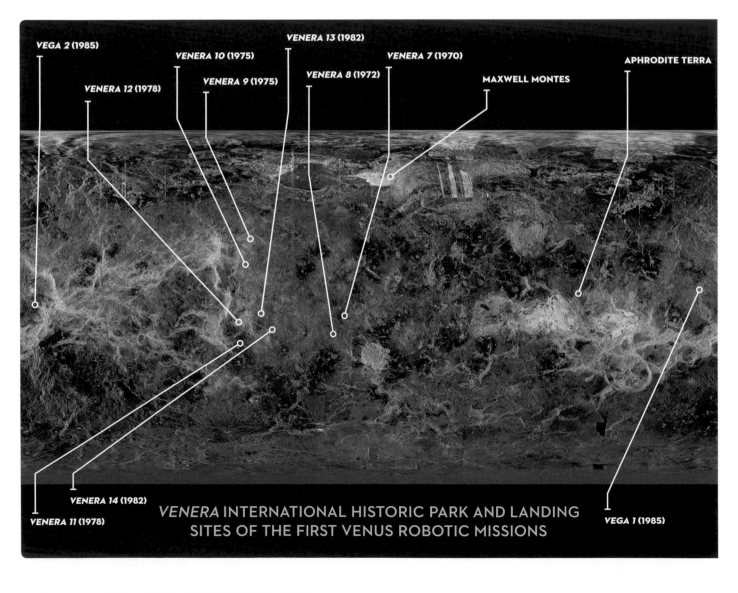

VEGA 2 (1985)
VENERA 13 (1982)
VENERA 10 (1975)
VENERA 12 (1978)
VENERA 9 (1975)
VENERA 8 (1972)
VENERA 7 (1970)
MAXWELL MONTES
APHRODITE TERRA
VENERA 14 (1982)
VENERA 11 (1978)
VEGA 1 (1985)

VENERA INTERNATIONAL HISTORIC PARK AND LANDING SITES OF THE FIRST VENUS ROBOTIC MISSIONS

VENUS FAST FACTS

Type of Body
Terrestrial (rocky) planet

✳

Distance from Sun
Average is ~0.72 AU, or 67 million mi.
(108 million km)

✳

Distance from Earth:
Ranges from 24 million to 162 million
mi. (39 million to 261 million km)

Travel Time from Earth
Varies from about 1 week to 100 days

✳

Diameter
7,521 miles (12,103 km), only about 5%
smaller than the Earth

✳

Highlight
Crazy hot surface, crazy high winds,
crazy acid clouds!

AVERAGE TEMPERATURES*

	DAYTIME OR NIGHTTIME	
	°F	°C
Average Surface	860	460

*The average surface temperature on Venus is the same day and night, from the equator to the poles!

Extremely High Temperatures: The overarching concern on Venus is what an engineer would call *thermal control.* How to keep habitats and suits and other systems—like people—cool on a super-hot world is a major task for the designers of Venus tour operations. If you want to go out onto the surface, you'll have to learn how to get into and operate a hotsuit, which is a sort of cross between an old Apollo-style space suit and a highly reflective metallic coat of armor made out of titanium. Alternatively, if you don't want to go outside, you can take a bumpy ride to the surface in a sturdy hotpod—a powered tram designed to carry up to 20 people in relative comfort. Its giant thermally protected windows allow you to see the sights without having to suit up. The best way to beat the heat, however, is to stay in one of the airhotels, which are suspended by giant balloons about 30 miles (50 kilometers) up in the atmosphere. There, the temperatures and pressures are relatively Earthlike, and you can sometimes wander outdoors wearing only an oxygen mask. Don't worry, you'll still be able to see the surface passing quickly below as your hotel zips along in the high winds.

High Pressure: Related to the high temperature is, of course, the crushing atmospheric pressure of the surface. Here, again, a hotsuit or hotpod is essential, as they are built like the uncrushable, super-strong submersibles that take tourists to the bottom of the Earth's deepest oceans, where the pressure is even greater than on Venus.

Acid Rain: As your transport vehicle coming from Earth descends into the atmosphere, you will have to pass through several layers of noxious, sulfuric acid clouds; fog; and even rain before you get down to the level of the airhotels. Your transport will, of course, be equipped to handle the corrosive acid bath (though you may get a whiff of rotten eggs now and then if the ventilation system is not working at 100 percent). Once at your airhotel, you should expect that there will be several times during your

THE GREENHOUSE EFFECT

Some atmospheric gases, such as water vapor and CO_2, allow visible light from the Sun to pass through them and reach a planet's surface; however, these gases then absorb prolific amounts of infrared (heat) energy that tries to pass back out from the surface to space. These kinds of gases are called greenhouse gases, which is analogous to the way greenhouse windows let in sunlight but prevent heat from getting back out. CO_2 is a very efficient greenhouse gas, and because Venus has so much of it in its atmosphere, it traps the outgoing heat quite effectively and causes the surface temperature to rise to crazy hot levels.

While massive greenhouse warming can wreak havoc on a planet's climate, a little greenhouse warming can be a good thing. Indeed, scientists have realized that Earth is a habitable oceanic world only because of the influence of two relatively minor but critically important atmospheric greenhouse gases: CO_2 and H_2O (water vapor). Without the 30 degrees Fahrenheit (17 degrees Celsius) or so of greenhouse-gas warming that they produce naturally in our atmosphere, Earth's oceans would freeze solid and life on our planet would likely be very different—if indeed any life had developed at all.

Scientists in the early twentieth century realized that past decreases in CO_2 over geologic time could explain the ice ages, and that continued burning of fossil fuels could enhance the CO_2 abundance and lead to global warming. In many ways, Venus has helped us to understand what the greenhouse effect is and how it works—and how the effects of greenhouse warming influence the climate of our own world.

stay when you'll have to batten down your windows and doors and stay inside during "acid drills"—times when large storms or turbulence in the sulfuric acid clouds bring the acid zone close to (or at) the altitude where the airhotels are floating. Typically the hotel operators will lower their altitude to help keep everyone safe, and the structures and balloons are built to survive exposure to acid. Still, it can be inconvenient, and getting caught in an acid storm is almost certain to be deadly.

DON'T MISS . . .

Venus is a world of extremes—heat, pressure, winds, acid rain, and not a hint of water in sight. But, like many of the deserts on Earth, the surface of Venus has a stark natural beauty that draws visitors willing to brave the extreme conditions. While the era of Venus tourism is still relatively

MAXWELL MONTES
VENUS

A fanciful artist rendering of the peak of Maxwell Montes.

A perspective rendition of Maxwell Montes, drawn from topographic and radar data.

people in the solar system who have witnessed firsthand the snows of Venus. Wait—snow on crazy hot Venus? Sort of. At the super-high elevation of Maxwell Montes, the temperatures and pressures are a bit lower than down in the plains, and so some of the minerals that are normally vaporized down in the hotter plains are stable near the summit. Among these "snowy" minerals is pyrite—an iron sulfide mineral commonly known as fool's gold—which glistens in the diffuse sunlight like snow at the top of a high mountain peak on Earth. It's a magical sight to behold, and if you're lucky (or if you've prepaid), the crew might let you go out for a walk in the crunchy pyrite snow using a hotsuit.

VENERA INTERNATIONAL HISTORIC PARK

📖

Between 1961 and 1984 the USSR launched an astounding 28 robotic exploration missions to Venus, including multiple flybys, orbiters, and 9 robotic landers that operated successfully on the surface—2 of which deployed scientific balloon experiments on the way down. It was the first golden age of Venus exploration, and (like on the Moon), the historic remains of these first visits to Venus are scattered across the surface at landing sites that have since become UNESCO Interplanetary Scientific Heritage Sites.

young, there are nonetheless already a few amazing sights and experiences that qualify as must-see destinations on your visit to Earth's fraternal twin. These include:

THE SNOWS OF MAXWELL MONTES

🏔️ 📷

Most of Venus's surface is made of dark, rolling plains of lava that erupted globally in a fit of volcanic activity some 500 million years ago—planetary scientists are still trying to explain it. But, in a few places, rising above those plains are some majestic mountains that dwarf the tallest peaks on Earth. The tallest of them all is Maxwell Montes (named after the nineteenth-century British physicist James Clerk Maxwell, who pioneered the study of electricity and magnetism). Located in Ishtar Terra and rising more than 36,000 feet (10,973 meters) above the plains, it dwarfs the Earth's tallest mountains (which peak at just under 30,000 feet [9,144 meters]). Cytherean Tours Inc. offers a spectacular all-day hotpod aerial tour along the mountain's lower slopes and flanks, eventually landing near the peak. There, you will be among the few

A view of the surface of Venus, obtained by the *Venera 13* lander.

WHERE ARE VENUS'S OCEANS?

Even though Venus is about the same size as Earth, and only just a little closer to the Sun, it's a bone-dry world—no liquid water, ice, or even detectable water vapor among its hotter-than-an-oven atmosphere of sulfuric acid clouds. On Earth, CO_2 and other gases released from volcanoes can't build up too quickly, because they dissolve in our ocean and are trapped within carbonate minerals in rocks. If Venus had an ocean, this would be true there as well. But without an ocean, the CO_2 and other gases continued to build up over time, creating a world that got hotter and hotter, boiling off any water that may have been

there earlier in the planet's history. Indeed, many scientists think that Venus may have had an ocean early in its history, based on the similarity of its initial circumstances to those of our own planet. If so, clearly something went horribly wrong. Maybe the ocean was lost in the same catastrophic event (perhaps a giant asteroid impact?) that some scientists believe might be responsible for the planet's extremely slow, backward spin. Or maybe the water slowly trickled away into space because the planet doesn't have a protective magnetic field like Earth does. Where are the oceans of Venus? It's a mystery.

Left The *Venera 7* spacecraft being assembled.

Right The egg-shaped *Venera 7* landing module.

It is perhaps no surprise, then, that short visits to some of those landing sites have become popular tourist activities on Venus. The egg-like *Venera 7* lander, for example, is the first human-made object to land on another planet (back in 1970), surviving and relaying scientific data for a mere 23 minutes in the extreme heat and pressure at the surface. Nonetheless, it is still there, serving as a lonely technological sentinel in an otherwise barren and unforgiving landscape, and you can take a hotpod tour to see it up close. Another popular destination is the landing site of the *Venera 13* lander, which, in 1982, radioed back

the first color photos from the surface of Venus, revealing a rugged, orange-brown landscape reminiscent of a cloudy day somewhere on Mars. Several tour operators offer trips to these and other *Venera* landing sites, and they also include detailed historical and scientific accounts of the missions and their main results.

HOTWING TOUR

A recent entrant in the "extreme solar system thrill-seeking" competition is a daring new aerial excursion now offered by several of the airhotels near Maxwell Montes. For a steep fee (and after signing the right insurance release forms), you can don a set of hotwings, plunge off the deck of the hotel, and fly like a bird through the thick air of the Venus atmosphere. Hotwings are a combination of a hotsuit and a set of winglike flaps that, when extended, provide lift just like the wings of a bird. Reminiscent of the old wingsuits that became popular for gliding down windy mountain slopes in the early twenty-first century, hotwings enable true

HISTORY OF EXPLORING VENUS

1032: Persian astronomer Abu Ali ibn Sina notices a spot moving across (transiting) the face of the Sun, possibly the first Venus transit observed by humans

1610: Galileo becomes the first person to observe Venus through a telescope

1639: Jeremiah Horrocks observes the first accurately predicted transit of Venus

1761: Venus discovered to have an atmosphere during solar transit observations by Russian astronomer Mikhail Lomonosov

1927: First photographs of Venus show bright and dark atmospheric markings

1932: First spectroscopic identification of CO_2 in Venus's atmosphere

1956: First radio-wavelength observations of Venus indicate very high surface temperature

1962: NASA's *Mariner 2* robotic spacecraft performs first flyby of Venus

1964: Venus's super-slow rotation rate discovered by Arecibo Observatory radar observations

1967: USSR's *Venera 4* robotic spacecraft becomes first successful probe to study Venus's atmosphere

1970: First Venus lander, the *Venera 7* robotic spacecraft, confirms super-high surface pressure and temperature

1972–85: Successful series of eight more USSR *Venera* and *Vega* robotic landers

1975: *Venera 9* becomes first robotic spacecraft to orbit Venus

1985: *Vega 1* robotic mission flies the first balloon experiments on another planet

1990: *Magellan* robotic spacecraft generates first global radar map of Venus

2030: First Venus robotic sample-return mission determines details of atmospheric composition

2056: First humans orbit Venus, remotely obtain new surface and atmospheric data

2141: First astronauts land on Venus

2175: First airborne research outpost, Aphrodite, established in Venus's stratosphere

2190: First tourist flights and first Cytherea airhotels established

2218: Ten-thousandth hotsuit tourist expected to trek across the surface of Venus

human-powered flight because of the thick atmosphere below the airhotels and the planet's high winds that help provide lift. It is a lot of work, however, and quite dangerous, because of the high turbulence in the atmosphere and the possibility of sudden acid storms that significantly reduce visibility and flying efficiency. While it is fascinating to watch some of the locals who live and work in the airhotels and have become stunningly good at flying, you'd better be in tip-top shape to take up the pastime on your own.

GETTING THERE

Several spacelines offer transport to Venus, though the flights are less frequent and of course much longer than voyages to the Moon. One-way travel times are typically around three months using shuttles with traditional propulsion systems, although newer-technology shuttles can cut that down to around a week, if you're willing to pay significantly higher prices. Part of the reason the new shuttles can get there so fast is that they use the friction of the atmosphere of Venus to slow down, a maneuver

CURRENT VENUS AIRHOTEL CHOICES

AIRHOTEL	LOCATION	TARGET	THE VIBE
Cytherea 1	Drifting along equator	First-timers	Oldest airhotel chain, classic late-twenty-second-century designs, great sampling of solar system cuisine
Cytherea 2			
FairWinds	Varies in latitude to avoid storms	Adventure types	Wide selection of bars and pubs, offers hotwings lessons, base for camping and hiking excursions
Aphrodite	Circles over Aphrodite Terra	Couples	Spectacular views of rugged mountains, best spa and yoga spots
Maxwell	Tethered near Maxwell Montes	Families	Best choice of kids' play and food options, offers Camp Venus day care

A high-altitude shuttle ferrying passengers to a Venus airhotel.

Several hotpod tours provide dramatic views of some of the active volcanoes of Venus.

known as aerobraking, which has been used on many robotic space missions since the 1970s. It's a crazy, rough, turbulent roller-coaster ride for passengers, though, so if you've got a weak stomach you might consider letting the crew sedate you ahead of time.

Once you're on Venus, each airhotel will have a small fleet of hotpods and other trams that can transport you to the surface or to other airhotels if you're looking to change things up during your stay.

THINGS TO DO, PLACES TO STAY

The current selection of airhotels scattered around the planet are the central attraction of Venusian tourism, providing accommodations, dining, in-house activities, and access to a variety of excursions on or near the surface. Some people never leave the comfort of their airhotel, preferring to experience the extremes of Venus remotely, drifting high above the surface in the stiff breezes. Others want to brave the surface, which they can do either for a few hours in a hotsuit or for longer durations on camping trips offered at a few hardy but spartan structures that have been built on the surface. "I went camping on Venus" is not something you hear people say very often around

the solar system, which makes some adventure travelers covet the experience even more.

Dining choices are mostly limited to imported foods, of course, but each airhotel will have a small selection of produce locally grown in their floating gardens, and all have top-notch 4- and 5-star dining options if you feel like splurging. On the other end of the spectrum, you can find standard-fare fast food in each airhotel, although FairWinds has become planet-renowned for having the best pub grub this side of Earth.

LOCAL FLAIR

The extreme environment of Venus attracts many extreme thrill seekers to work at the airhotels and excursion companies. If hotwing lessons don't push you hard enough, there are apparently even more dangerous extreme sports activities that only the locals partake in so far. For example, there are rumors that some of the local teenagers put one another through a lava-rafting initiation rite that takes place along some of the recently active volcanic flows near Maat Mons. Whether it's true or not, you'll still have to spend some time getting to know the locals, and getting them to know and trust you, before they'd be likely to reveal any of the "hidden Venus" secrets that could make your visit the trip of a lifetime.

VISIT
Mercury
AND THE CALORIS BASIN

ZIPPING AROUND ON MERCURY

You'll quickly learn why Mercury was named after the fleet-footed messenger of the gods as you gain speed on your approach to the solar system's inner-most planet. Orbiting at only about a third of Earth's distance from the Sun, Mercury is much deeper inside what astronomers call the Sun's gravity well. The increased gravitational force from the Sun speeds up the planet, which takes just 88 days to go around our star. And oh, does that star loom large in the Mercurian sky, appearing about two and a half times larger than it does on Earth! You'll need to prepare for more than just extra sunscreen, however, as the effects of the Sun's proximity also include much higher daytime temperatures on spacecraft, space suits, and habitats. Thermal control is job number one for Mercury engineers and tour guides.

Opposite: The 960-mile-wide (1,550-kilometer) Caloris impact basin dominates the geology of nearly an entire hemisphere of Mercury. (Illustration by Lynx Art Collection)

Mercury's high-speed race around the Sun has attracted racers in general to the innermost planet. Once you land on Mercury, you're already traveling four times faster around the Sun than you were on Earth, which has allowed shuttle-speeder pilots to set records year after year for the fastest sprints around Mercurian race courses, the planet, or even around the Sun. Mercury tour companies attract racing and rocketry motorheads from all over the solar system—some who race, others who tinker with technology, and some who just want a dream-come-true experience of getting up close and personal with the cutting edge of racing speed and agility.

BEFORE YOU GO

At first glance, Mercury might seem a lot like the Moon. In some ways it is—a relatively small gray world with no atmosphere and lots of impact craters. But in other ways it is dramatically different, with more gravity than the Moon, harsher temperatures, and a much less-developed tourism industry. Mercury has yet to undergo a true boom in tourism, and so your choices are quite limited. Still, for your various Mercurian explorations, you'll want to be prepared for the following:

Extreme **Temperatures:** Because it is so close to the Sun, daytime high temperatures on Mercury can reach upward of 800 degrees Fahrenheit (425 degrees Celsius), and because there is no atmosphere or ocean to

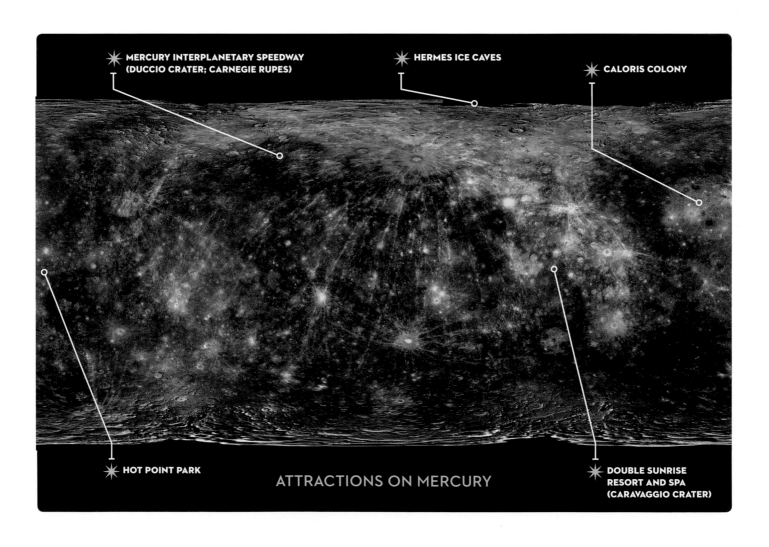

MERCURY INTERPLANETARY SPEEDWAY
(DUCCIO CRATER; CARNEGIE RUPES)

HERMES ICE CAVES

CALORIS COLONY

HOT POINT PARK

ATTRACTIONS ON MERCURY

DOUBLE SUNRISE
RESORT AND SPA
(CARAVAGGIO CRATER)

MERCURY FAST FACTS

Type of Body
Rocky/metallic planet, just 5.5% of the mass of the Earth

✳

Distance from Sun
Average is ~0.39 AU, or 36 million mi. (58 million km)

✳

Distance from Earth:
Average ranges from 48 million to 138 million mi. (77 million to 222 million km)

Travel time from Earth
Varies from ~10 days to 120 days

✳

Diameter
3,032 mi. (4,880 km), about 38% of Earth's diameter

✳

Highlight
Super hot or super cold, but always moving fast!

AVERAGE TEMPERATURES

SURFACE AVERAGE		DAYTIME HIGH		NIGHTTIME LOW	
°F	°C	°F	°C	°F	°C
152	67	800	425	-328	-200

help keep temperatures steady, nighttime temperatures on Mercury can plunge to nearly -328 degrees Fahrenheit (-200 degrees Celsius). This huge day-to-night temperature swing is the largest in the solar system, and it puts enormous stresses on spacecraft, space suits, and other systems on and around Mercury, as metals and other materials expand in the heat and shrink in the cold. Luckily, many of these thermal expansion problems were solved in the early days of lunar colonization. Even though the day-to-night temperature swing on the Moon is only about half of that on Mercury, figuring out the basic physics and engineering for thermal control on the Moon allowed engineers to expand their designs to Mercury's environment. The resulting ships and habitats are a little more insulated, and the suits a little bulkier, but once you've figured out how to use them (and fully appreciate how they keep you alive), you'll get on just fine.

Unexpected Gravity: If you've spent a lot of time on the Moon, you might take a bit longer to adapt to Mercury's unexpectedly higher gravity, which is about 2.3 times greater than that on the Moon. You'll have to adopt a new gait if you're out and about on the surface of Mercury. Luckily, it will be closer to your walking style on Earth—although the gravity on Mercury is still nearly 40 percent less than on Earth.

Extremely Rough Terrain: Like the bright highlands of the Moon, Mercury's surface is heavily cratered, rough, and littered with boulders and other potential hazards. Mercury's roughness, however, is more relentless, since it does not have as many of the large, smooth, ancient lava plains that we recognize as the "face" of the Man on the Moon. Walk with care, as a trip can lead to a tear in your space suit!

Mercury—shown here in an accurate size comparison to the Earth (left) and Moon (right)—is the smallest of the terrestrial planets. Its position as closest to the Sun also makes it the fastest-orbiting planet in our solar system.

Strange Days and Nights: Mercury's strange and unique orbit around the Sun will wreak havoc with your normal Earthbound circadian rhythms, as it takes about 88 Earth days for the Sun to make its way across the sky from sunrise to sunset, and the Sun even changes direction a few times while traveling along its course. Some of the planet's accommodations ignore this strange cycle, shut out the Sun, and maintain Earthlike 24-hour light-to-dark cycles inside. If you're out and about, however, you might get confused, thrilled, or both, by the odd timing of day and night.

DON'T MISS . . .

While Mercury does not yet have the tourist infrastructure of the Moon or Mars, there are nonetheless a variety of unique and fun sights and activities that will make your stay on the First Planet memorable. These include:

MERCURY INTERPLANETARY SPEEDWAY

Speaking of sports, speed racing is king of the Mercurian sporting world, with a huge variety of classes of spaceships competing for highest velocity bragging rights. One of the most fun and exciting places to see the races (as well as the racers and their equipment) up close is at the Mercury Interplanetary Speedway, built along part of an impressive mile-high tectonic cliff called Carnegie Rupes that crosses through Duccio crater. Racers test their agility in wheeled and flying speeders by trying to stay close to the cliff walls while turning sharp corners or spiraling through several of the smaller impact craters within Duccio. It is a dangerous, high-speed thrill ride for racers and spectators alike. If it's speed and agility you're after, you won't find higher-tech equipment or more-revved-up racers anywhere else in the solar system.

WHY IS A MERCURIAN DAY TWO MERCURIAN YEARS LONG?

Unlike any other planet in the solar system, Mercury spins on its axis three times for every two orbits it makes around the Sun. Early astronomers had thought that perhaps Mercury would spin once on its axis for each single orbit around the Sun, like the Moon spins around the Earth. Such a tidally locked orbit is typical of very small worlds in orbit around much larger bodies. But that 1:1 orbit (one spin for one orbit around the Sun) only works for planets or moons that travel in basically circular orbits around their master. Mercury's weird 3:2 spin-orbit ratio is the result of Mercury's orbit being far from circular.

In fact, it is the most elliptical, or egg-shaped, orbit around the Sun of all the planets, changing its distance to the Sun by nearly 40 percent over the course of its "year." Because Mercury moves much faster in its orbit when it is closest to the Sun than when it is farthest away, it can't stay in a simple 1:1 spin state like the Moon. And because the apparent movement of the Sun across the sky is a combination of the planet's spin *and* its fast motion around the Sun itself, it takes two full orbits around the Sun to go from one sunrise to the next. Strange days indeed.

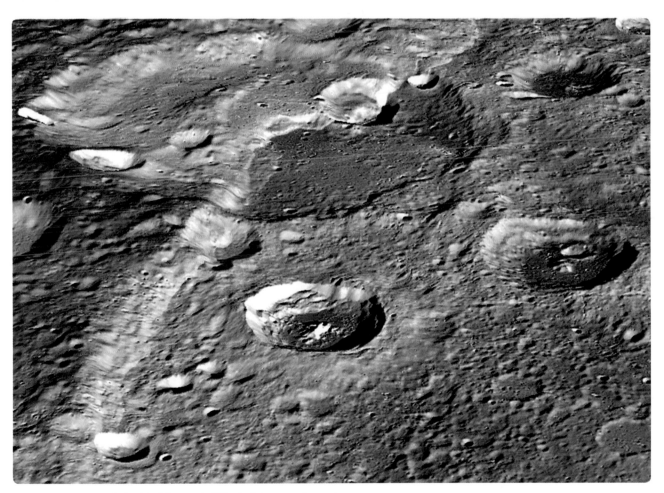

An enormous 1.2-mile-high (2-kilometer) cliff called Carnegie Rupes, here seen cutting through the 65-mile-wide (105-kilometer) Duccio crater, is the site of one of Mercury's most famous speedway courses.

THE INCREDIBLE SHRINKING PLANET

Mercury is crisscrossed with hundreds of giant cliffs like those at Carnegie Rupes—but why? These giant cliffs are tectonic features called thrust faults, which form on rocky planets that are slowly cooling over time. As the rocks cool, they shrink slightly, causing enormous compressional forces that can fold and tear a planet's crust. Planetary scientists think that Mercury has shrunk by perhaps as many as 20 miles (10 kilometers) in diameter since it formed, and the resulting thrust faults have created a global network of cliffs up to 620 miles (1,000 kilometers) long and 1.9 miles (3 kilometers) high. Similar thrust faults on the Moon indicate that it has slowly shrunk as it has cooled. On the Earth, plate tectonics and a still partially molten interior prevent the formation of such enormous worldwide cliffs.

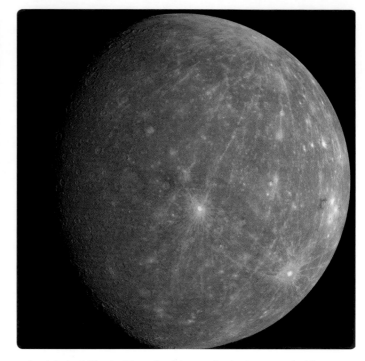

It might look like the Moon, but Mercury is a lot hotter, and while there you'll weigh more than twice what you would on the Moon (though still only about a third of what you weigh on Earth).

CALORIS COLONY

Your transport will arrive at the only major base built so far on Mercury: Caloris Colony, located within the enormous Caloris impact basin. At about 960 miles (1,550 kilometers) in diameter, Caloris is one of the largest impact craters in the solar system. The base itself is near the center of the basin, and in addition to the spaceport, it houses the largest concentration of hotels and restaurants on the planet. Accommodations are relatively basic, however, and the dining choices lean mostly toward (perhaps appropriately) fast-food options, apparently because most visitors to Mercury spend the majority of their time in off-base race-related activities, using the base itself only as a respite between races or excursions. Still, even if your Mercurian travels never take you beyond Caloris Colony, you can find good options there for local excursions to the relatively smooth plains nearby, as well as excellent real-time coverage of the seasonal speeder races shown on big screens or at holographic facilities in the many popular sports bars.

DOUBLE SUNRISE RESORT AND SPA

One of the most beautiful and fascinating of the natural phenomena that result from Mercury's strange 3:2 spin-orbit ratio—and its highly elliptical orbit around the Sun—is that people on Mercury can occasionally (but predictably) observe the planet's now-famous "double sunrises" from special places and at special times. One of these special places is the Double Sunrise Resort and Spa, located along the equator near 270 degrees west longitude (near the crater Caravaggio). From here, the Sun starts rising in the east just a few days before Mercury is closest to the Sun in its elliptical orbit, a position known as periapse. The planet's orbital velocity speeds up approaching periapse, eventually exceeding its spin velocity, making the Sun's motion in the sky stop and reverse, so that the Sun *sets* in the east! But then as Mercury's orbital velocity starts slowing down away from periapse, and its spin velocity once again exceeds the orbital velocity, the Sun goes back to its normal motion, rising *again* in the east a few days later and then moving normally across the sky. It all happens in slow motion, over the course of a few weeks, a couple of times every Earth year. The resort and spa offers spectacular east-facing, full-window cabins and other venues where you can relax, read, rejuvenate, and watch a celestial drama unfold unlike anywhere else.

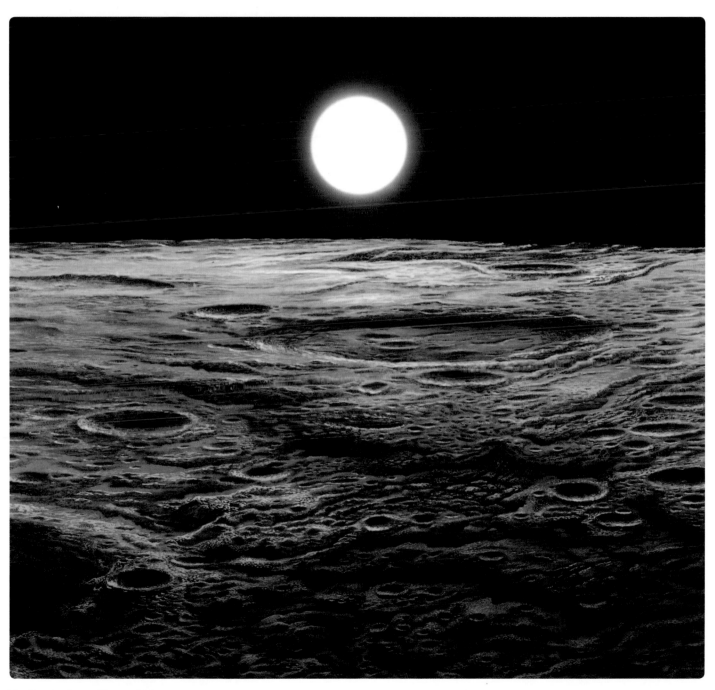

The Sun dominates Mercury's sky.

HOT POINT PARK

There are also places on Mercury where the Sun remains high overhead for extra-long periods of time, bobbing back and forth high in the sky, instead of along the horizon, when Mercury passes close to the Sun. The combination of "close to the Sun" and "Sun high in the sky" results in higher surface temperatures in these areas at these times. At one of these places, near the equator and around 180 degrees longitude, several local tour companies will take you out to a relatively smooth equatorial plain of ancient lava, which they're informally calling Hot Point Park, where you can get out and walk around (suitably equipped with a hotsuit) on the 800-degrees-Fahrenheit (425-degrees-Celsius) rocks. Even though the surface of Venus is just slightly hotter, you'd still have bragging rights to standing on the *second hottest* planetary surface in the solar system—and without having to deal with sulfuric acid clouds or bone-crushing atmospheric pressure.

HERMES ICE CAVES

Like the Moon, it was discovered back in the early space age that Mercury has permanently shadowed craters near its north and south poles (see box on page 8). These craters have since been thoroughly explored by planetary geologists, revealing a rich collection of ices brought in from the outer solar system by asteroid and comet impacts. Indeed, the water ice found at the poles is now commercially mined to support the water and oxygen needs of the colonists and tourists, just like on the Moon. While mine tours are not yet offered on Mercury, one of the ice-rich north polar craters, Kandinsky (craters on Mercury are named after famous writers, artists, and poets—in this case the Russian painter Wassily Kandinsky), houses a wonderful network of deep ice caves that has become a popular tourist destination. Enormous and sometimes bizarre crystals of ices made out of water, carbon dioxide, nitrogen, and other exotic compounds line the walls, ceilings, and floors of parts of the cave

system. A comfortable (well-heated) tramway has been built to take tourists along a spectacularly lit 6-mile (10-kilometer) path through the thickest of the crystals, which the mining consortium has agreed to set aside as a natural preserve. It is a spectacular and breathtaking sight to behold—and unexpected in one of the solar system's hottest worlds.

GETTING THERE

Getting to Mercury can be a challenge unless you're planning to go during "speed season," the several-week period that occurs about four times each Earth year, when Mercury is near its closest—and thus fastest—point in its orbit around the Sun. During these times, all of the major spacelines offer special charter excursions that bring thousands of racers, crew, and spectators to quickly fill up the planet's relatively few orbiting and surface hotels and resorts. Booking ahead is a requirement, especially for every fourth speed season, when the Solar System 500 race will determine the newest speed kings and queens to rule for the next Earth year.

If you're not able to go during speed season—or if you just want to avoid the crowds—you can still book passage on one of the several supply vessels that make the Earth–Mercury run every month or so. You won't travel first class, and it could take up to a few months to get there, but you'll find the costs quite reasonable. The cargo carriers also offer package deals with Mercury's resorts and hotels, as they are often relatively underbooked outside of speed season.

THINGS TO DO, PLACES TO STAY

Mercurian tourism has been built around speed season. Races can be viewed from orbital platforms, special holographic viewing centers in hotels and resorts at Caloris Colony, or modest accommodations at or near the raceways. Rooms and amenities range from spartan to 5-star, but all need to be reserved early to accommodate the speed-season crowds. In addition to seeing the races, you'll have the opportunity to take

HISTORY OF EXPLORING MERCURY

1610: Galileo becomes the first person to observe Mercury through the telescope

1630: Johannes Kepler accurately predicts the once-per-decade passage of Mercury's disk across the face of the Sun, not observed by astronomers until 1631

1928: First photographs of Mercury show bright and dark surface markings

1973: NASA's *Mariner 10* becomes first robotic spacecraft to fly past Mercury

1991: Ice discovered in Mercury's dark polar craters using Earth-based radar

2004: NASA's *MESSENGER* robotic spacecraft becomes first Mercury orbiter

2018: Joint European and Japanese *BepiColombo* spacecraft orbits Mercury

2054: First Mercury lander directly samples polar ice deposits

2075: First human explorers land on Mercury, return samples of polar ices

2110: Caloris Colony established

2151: First tourist flights and hotels established

2160: First organized land-speeder races (at the old Mercury Speedway)

2194: New Mercury Interplanetary Speedway opens for rocket-powered speeders

2218: Solar system speed championship races

pit tours of many of the vehicles (especially the ones that berth at Caloris), and even to race on some of the courses in slower (and safer) speeders, either before or after the main races.

If you're not into speed season, there are still some wonderful ways to explore the innermost planet and its unique environment, whether out hiking on the heat of the mid-day trails around Hot Point Park, cooling off in the underground ice cave tramway near the north pole, or enjoying spectacular surface or orbital views of our enormous, looming star dominating the Mercurian skies. Hotels are more affordable and rooms are typically available outside speed season.

LOCAL FLAIR

During speed season, Mercury is crowded, expensive, and nonstop. It can be exhilarating but also exhausting for the locals who work in and support the racing industry. If you're able to visit the innermost planet outside of speed season, you'll have a much better chance of getting to know some of the locals, or at least getting to catch them when they aren't stressed-out and overworked. Some visitors have even reported that off-season friendships struck up with Mercurian dwellers have led to spectacular "insider" access to private tours of racing vehicles and raceways, and even invitations to dinners or parties with some of the racers themselves (many of whom stay on-planet during the off-season). Be sensitive and understanding with the locals about the challenges of accommodating visitors like you during speed season. It just might pay off when the races are done.

MARS
MULTIPLE TOURS AVAILABLE
ROBOTIC PIONEERS / ARTS & CULTURE / ARCHITECTURE & AGRICULTURE

SUMMER VACATION ON MARS!

Mars is the solar system's most popular and visited deep-space tourist destination. Orbiting beyond the Earth and the Moon, the planet is a cornucopia of fantastic geologic, atmospheric, and polar features and phenomena. Activities abound for hikers, campers, foodies, photographers, writers, artists, sports fans, and followers of space-exploration history. Don't even think about trying to squeeze a visit to this, the most Earthlike of our fellow planets, into a weekend or even a week—take some extra time off work, pack up the kids, and spend the summer exploring all that the Red Planet has to offer!

Mars has held a special place in human history going back to prehistoric times. Its approximately biannual appearance as a bright, reddish "star" was among the earliest clues that the night sky is not static, but rather a dynamic canvas upon which, to our ancient ancestors, the actions of the gods were playing out.

Opposite: Mars has been visited by orbiters, landers, and rovers since the 1970s, and by people since the mid-twenty-first century. Many of those early historic sites have been carefully restored or preserved, so be sure to visit them on your trip to the Red Planet.

To the Greeks, the red color of Mars signified blood, and they named the wandering star after their god of war, Ares—later renamed Mars by the Romans. Many centuries later, Mars's distinctive motion across the night sky (part of which involves apparent backward, or retrograde, motion) helped convince astronomers of the early Enlightenment that the planets orbited the Sun instead of the Earth, and that many (including Mars) orbited in eccentric paths rather than in perfect circles.

Mars remained enigmatic right up to the beginning of the space age, a subject of fantasy and conjecture for science-fiction novelists and screenwriters who asked, "So what is Mars really like?" In the late twentieth and early twenty-first centuries, robotic flyby, orbiter, lander, and rover missions revealed a landscape that at once seemed familiar—akin to the American southwest or the Dry Valleys of Antarctica—but at the same time was more dangerous and hostile to human life than Earth's most extreme cold or desert environments.

Compared to Earth, Mars has only three-eighths its gravity, so you'll be able to walk, run, and jump on Mars with relative ease. It's typically extremely cold on the surface, with average daytime and nighttime temperatures over the course of a Mars year of -63 degrees Fahrenheit (-53 degrees Celsius) and -135 degrees Fahrenheit (-93 degrees Celsius), respectively. And, by the way, while there's often a pretty strong breeze, the air is super thin (only 1 percent of the pressure

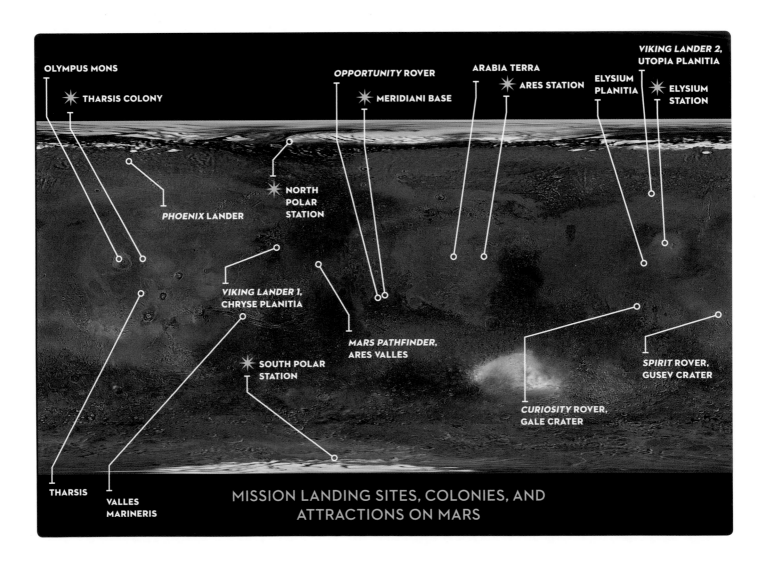

OLYMPUS MONS

THARSIS COLONY

OPPORTUNITY ROVER

MERIDIANI BASE

ARABIA TERRA

ARES STATION

ELYSIUM PLANITIA

VIKING LANDER 2, UTOPIA PLANITIA

ELYSIUM STATION

PHOENIX LANDER

NORTH POLAR STATION

VIKING LANDER 1, CHRYSE PLANITIA

MARS PATHFINDER, ARES VALLES

SOUTH POLAR STATION

CURIOSITY ROVER, GALE CRATER

SPIRIT ROVER, GUSEV CRATER

THARSIS

VALLES MARINERIS

MISSION LANDING SITES, COLONIES, AND
ATTRACTIONS ON MARS

MARS FAST FACTS

Type of body
Planet

✳

Distance from Sun
Averages 1.52 AU, or 142 million mi.
(228 million km), though the planet's
eccentric orbit makes the actual solar
distance vary from 1.38 to 1.66 AU

✳

Distance from Earth
Average Varies from 34 million mi.
(55 million km) to 248 million mi.
(400 million km)

Travel time from Earth
Varies from ~2 weeks to
~6–9 months

✳

Diameter
4,220 mi. (6,792 km), about 53% of
Earth's diameter

✳

Highlight
Tallest volcanoes! Deepest canyons!
Craziest dust storms! It's Earth
gone wild!

AVERAGE TEMPERATURES

	DAYTIME HIGH		NIGHTTIME LOW	
	°F	°C	°F	°C
Yearly Average	-63	-53	-135	-93
Equatorial	70	21	-54	-48
Midlatitude	-45	-43	-153	-103
Polar Day	-9	-23		
Polar Night			-198	-128

at Earth's surface) and unbreathable because it's 95 percent carbon dioxide (CO_2). While Mars's diameter is only half that of Earth's, the surface area is about the same as the land area of all the continents on Earth, and so if you're properly prepared and equipped, there's lots of room to explore!

BEFORE YOU GO

Perhaps the most dangerous thing about being on Mars is that it is easy to believe you're actually in an Earthlike environment. Many places *look* like Earth's desert environments, and the presence of habitats, vehicles, and people on the surface only adds to the illusion. Complacency about the cold, low pressure, hyper-low humidity, lack of oxygen, higher radiation levels, or the hazards of the ever-present dust is a recipe for disaster. Educate yourself about the distinctive hazards of the Martian environment, and be respectful of all the ways that this spectacularly beautiful world is trying to kill you.

During your various adventures on Mars, you'll want to be prepared for the following:

Low Temperatures: Prepare to be cold. That is, prepare to don appropriate thermal wear and the right kind of spacesuit for the locale you'll be visiting (equatorial, high altitude, polar, or subterranean). Tour companies can provide excellent gear for any activity, but remember that it can take days to get it fitted and learn to use it properly. Before you land on Mars, take advantage of opportunities to become proficient in the kinds of spacesuit technology you'll be using to save yourself a lot of hassle.

High Radiation: The atmosphere of Mars does not have an ozone layer that blocks high-energy ultraviolet radiation from the Sun like the Earth's atmosphere does. Likewise, Mars does not have a magnetic field to block or deflect the solar wind and other higher-energy forms of radiation. The result is that if you don't take precautions such as proper shielding of space suits, surface vehicles, and habitats, you could be subjected to dangerous levels of radiation over time, leading to sickness and other ailments. Be sure to use the equipment provided by your flight crew and tour guides to keep track of your total radiation dose while visiting.

Midlevel Gravity: The gravity on Mars is about three-eighths of what it is on Earth. So you'll feel lighter on Mars than you do on Earth, but not as light as you'd feel on the Moon—or on Phobos, Deimos, or other much smaller worlds. The moderately lower gravity can play tricks on you, giving your inner ear a false sense of Earthlike balance and even causing some disorientation and "sea sickness" in some people. As with all dramatic environmental changes, give your system time to adapt. Some people use weight suits on arrival, to simulate being in Earthlike gravity, and slowly remove weights as the days go by. Others just wear lightweight Mars suits and tolerate a little physical discomfort in the interest of adapting as quickly as possible.

Strange Terrain: Mars is a world covered in sharp volcanic rocks, shards of volcanic glass, loose sand, rough impact craters, and lava flows. Yet some places on Mars are quite smooth—and quite dangerous. Specifically, places where thick piles of the famous Martian dust have built up appear deceptively smooth and easy to navigate. But beware: If you step into a dust-filled crater or crevasse, for example, you might instantly plunge up to six feet (two meters) deep into the dust, and land on sharp rocks below. These fluffy terrains are sort of the Martian equivalent of quicksand back on Earth. One of the early Mars rovers, *Spirit*, succumbed to just this kind of fate while driving across Mars in the early twenty-first century, falling about 10 inches (25 centimeters) into an unrecognized dust-filled crater, from which it was unable to emerge. Luckily, your trail guides and park rangers will know how to recognize these places, and with a little rope will help you get out. Those new to Mars will want to arrange for special, extra-padded suits to guard against the rough terrain—and stay on the trails to avoid embarrassing and often painful plummets.

Dust, Dust, Dust! When you go to Mars, even if you stay inside the whole time, you're going to get dusty and dirty. The famous reddish dust of Mars—the ground-down bits of rock created by billions of years of impacts and erosion—has a consistency ranging between flour and smoke. Every time an airlock opens, dust somehow sneaks inside, despite the filtration systems. It gets into space suits, into hotel rooms, into the kitchen, into the food. You'll be eating Mars without even realizing it. Luckily, the Red Planet isn't toxic. But there *are* dangers due to dust, including damage to space suits, airlock seals, and mechanical equipment. Early Mars settlers had to deal with significant dust-related health problems, especially red lung, a respiratory-system ailment akin to the coal-dust-induced problem known as black lung, which was faced by miners for centuries back on Earth. Modern medicine and better filtration systems have eliminated the worst red-lung issues, but many visitors to Mars still experience scratchy, dry throats, and general irritation to the eyes, nose, and skin from the insidious dust. Shower often, drink lots of water, and pay attention to the safety drills.

THE CANALS OF MARS

In the early twentieth century, it was widely accepted by the public that Mars was home to a technologically advanced civilization capable of stunning engineering achievements—such as a global system of canals that could transport water from the poles to equatorial cities. Businessman and amateur astronomer Percival Lowell popularized this hypothesis with his telescopic drawings of Mars that showed fantastic networks of straight, interconnected lines that he assumed were enormous water-filled canals. Later, telescopic observations and space probes would dispel this popular notion, revealing instead an ancient, desiccated landscape upon which water did once flow, but billions of years ago. Perhaps ironically, in the early twenty-second century, part of the job of the newly constructed ice-mining companies was to design and build enormous, straight, interconnected series of pipelines—canals of sorts—to transport water from the poles to the colonies in the equatorial cities. Channeling twentieth-century science-fiction writer Ray Bradbury's prescient *The Martian Chronicles*, we have indeed become the Martians.

A view of the Valles Marineris (aka the Grand Canyon of Mars) hemisphere, taken on the late-twentieth-century robotic *Viking Orbiter* mission.

DON'T MISS . . .

You could spend a lifetime on the Red Planet and still not visit all the spectacular geologic wonders that the planet has to offer. Fortunately, many of the "greatest hits" of Mars have been turned into parks, making them widely accessible as popular destinations for major tour and excursion companies. Other attractions include uniquely Mars-themed resorts, restaurants, and entertainment venues, most of which are associated with one of the half dozen colonies and spaceports that have thus far been established on Mars. Some highlights include:

VALLES MARINERIS: FLY THROUGH THE "GRAND CANYON" OF MARS

Almost an entire hemisphere of Mars is dominated geologically by a huge rip in the crust, a gigantic chasm formed billions of years ago by the upwelling of hot subsurface volcanic magma and the ensuing rifting of the surface. This enormous valley, first discovered in images by the *Mariner 9* spacecraft back in the 1970s, is called Valles Marineris, or sometimes the Grand Canyon of Mars. In truth, however, it dwarfs the actual Grand Canyon back on Earth, as Valles Marineris stretches over 2,500 miles (4,000 kilometers) across the surface. If it were on Earth, the Grand Canyon of Mars would go from New York to Los Angeles. The canyon is more than 120 miles (200 kilometers) wide in places, and up to 23,000 feet (7 kilometers) deep! Layers in the walls of the canyon reveal ancient lake deposits and more recent lava flows, and the floor of the canyon has its own unique weather

Above: A perspective view of part of the walls and floor of the Valles Marineris canyon system.

Below left: Approaching the 85,000-foot-tall (26,000-meter) Olympus Mons, the largest volcano in the solar system.

patterns, including fog, haze, and dust storms. Numerous side canyons and valleys connect to the main span, providing a maze of geologic features to discover.

Numerous tour companies offer spectacular fly-throughs of the canyon, and many offer surface excursions along the floor and walls. You can easily spend a week or more exploring the geology and scenery, either by flight, foot, or both. In-canyon accommodations include campgrounds, simple dome habitats, and several high-end spa and natural bath resorts deep within the canyon that tap into the weak geothermal heat still rising from the subsurface.

EXPLORE THE VOLCANIC PEAKS OF THARSIS

The Tharsis region provides great hiking opportunities on four large volcanoes that cover an area about the size of the western half of the continental United States.

WHY ARE THERE SUCH GIANT VOLCANOES ON MARS?

Mars is a smaller planet than Earth, so why does it have much bigger volcanoes? Well, the answer comes from two factors: tectonics and gravity. Tectonics matters because, unlike Earth, Mars does not have a global set of crustal plates that move relative to one another. The lack of plate tectonics means that when a volcano erupts, lava from deep in the mantle of Mars piles up on the crust in one spot—higher and higher—rather than spreading out into chains of "hot spot" volcanoes like the Hawaiian Islands. Gravity matters because the lower gravity of Mars allows volcanic structures to rise to much higher heights before reaching equilibrium. On Earth, Olympus Mons would be at best a 30,000-foot-tall mountain— impressive for sure. But on Mars, the same kinds of volcanic processes, under lower gravity and without plate tectonics, led to the tallest mountains in the solar system.

SUMMIT

OLYMPUS MONS
The solar system's highest peak

The largest of them all, Olympus Mons, a single volcanic mountain with a base the size of Arizona, towers more than 85,000 feet (26 kilometers) above the surrounding lava plains. Now part of the Interplanetary Park System, Olympus Mons boasts dozens of trails and camping locations. Luckily, most of the trails are not steep (the mountain has gentle slopes shaped like a warrior's shield, typical of what geologists aptly call a shield volcano), but some parts are quite rugged, littered with the eroded and impacted debris of lava flows that are hundreds of millions of years old. To ascend those areas, you can ride a spectacular cable car up through some of the most rugged and dramatic scenery on the planet. While the volcano itself is extinct, geologists have discovered a few places where geothermal (actually, aerothermal) heat still makes its way to the surface, supporting a few small, human-made thermal hot-springs resorts that are great places to rest along the way to the top. When you get to the summit, sign the historic register, enjoy the view of the curvature of the planet through the thin atmosphere, and bask in the knowledge that you've literally reached the greatest heights possible on any planet in the solar system!

A spectacular cable car takes you up through magnificent vistas and the rugged terrain on the solar system's largest volcano, Olympus Mons.

SNOW ON MARS?

Mars is tilted on its axis about the same amount as the Earth is, and thus it has seasons like the Earth's—winter, spring, summer, and fall. Each season is about twice as long as back on Earth, however, because it takes Mars about twice as long to orbit the Sun. One effect of the seasons is that, like on Earth, both polar regions go through long periods of darkness—polar night—and then long periods of continuous daylight—the midnight sun. The polar regions are cold, and both poles harbor relatively small but persistent deposits of surface water ice, known as the perennial polar caps. During the polar night, temperatures drop dramatically, and when they hit about -198 degrees Fahrenheit (-128 degrees Celsius), it becomes so cold that CO_2, the dominant gas in the atmosphere, literally snows onto the ground as dry ice. This "snow" creates seasonal dry ice deposits that are up to 3 feet (1 meter) thick and often cover the perennial polar caps with water ice. When the Sun returns in the spring, they evaporate back into the atmosphere. This cycle of seasonal polar-cap growth and shrinkage has been monitored from Earth-based telescopes and Mars-orbiting spacecraft for centuries.

FUN IN THE POLAR SNOW

The north and south polar caps of Mars are winter-sports wonderlands for athletes and tourists. Permanent water-ice polar caps provide opportunities for some dramatic hiking, snowshoeing, snowmobiling, and even sledding and skiing. (You will need to buy or borrow special Mars-specific heated-bottom skis or sleds that manually create the low-friction surface conditions that naturally occur on the bottom of typical Earth skis and sleds.) For a truly unique and bizarre experience, also try to get out onto a seasonal polar cap while its dry ice is forming in the late fall, or before it evaporates in the early spring. Parts of the seasonal ice caps are eerily transparent; it will seem like you are walking on crystal-clear water or floating many feet above the actual surface as you walk. Other parts of the CO_2 caps have some of the most rugged, difficult topography you're ever likely to (attempt) to hike across. And in some places, you can watch spectacular water geysers erupt through the cap, as the Sun heats the icy surface below and causes the CO_2 ice to crack, violently releasing the trapped water vapor.

Resorts and hotels at both poles have been built into the ice and snow, including several that are deep enough to enable breathable atmospheres to be maintained (still, bring a sweater!).

ARES STATION COLONY

The first settlers on Mars needed to find a place to live where they could balance all the competing factors that were (and still are) dangerous risks to the survival of humans. These factors include extremely cold temperatures (thus the desire to stay close to the equator, where it's warmest), the need for ample solar power (again, favoring the equatorial region), the need for safety from high radiation levels (which can be found either in underground locations or in places where the soil/regolith can be worked and built in ways amenable to radiation protection), and proximity to a sufficient supply of surface or subsurface water. That final requirement tipped the scales for the early settlers toward northeastern Arabia Terra, one of the closest regions to the equator where subsurface ground ice as well as water-bearing mineral

An oblique view of the layered deposits at Mars's northern pole: snow, dust, and lots of fun!

deposits had been identified by early-twenty-first century robotic missions, and then confirmed by some of the first astronaut missions to Mars.

As the first human colony on Mars, Ares Station is rich with the history of space science and exploration. While it is small in comparison to lunar colonies and the newer Mars colonies, there's still a great selection of hotel accommodations and dining options, many of which include beautiful views of the surrounding crater rims and layered cliffs from the windows that protrude from the mostly buried base's primary structures.

THARSIS AND ELYSIUM LAVA TUBE COLONIES

All those volcanoes on Mars have done more than just create large mountains and surface lava flows. Many have also led to the creation of underground lava tubes, which molten lava once flowed through on its way to breaking out onto the surface somewhere downstream. When the lava supply dwindled and the molten rock cooled and solidified, these tubes became vast networks of caves and tunnels. They were first discovered in the early twenty-first century, when the presence of "skylight" holes, created by small impacts into the tops of some of these lava-tube caves, was noticed.

The most dramatic of these underground lava-tube networks in the Tharsis and Elysium volcanic regions were converted into habitats by some of the earliest Martian colonists, partly to help keep them shielded from dangerous levels of UV and cosmic radiation at the surface. Since then, Tharsis and Elysium have become full-fledged subterranean colonies. Now sealed off completely from the outside environment, they are heated to Earth shirtsleeve temperatures and filled with breathable air. Tourist hotels and resorts, as well as some of the best restaurants on the Red Planet (many of which serve food that has been grown or raised in the tubes themselves), now occupy special sectors of these colonies. Tourist-friendly activities there include museums, plays, concerts, casinos, and some sporting events.

THE FIRST HUMAN LANDING ON MARS

After the incredible successes of the Apollo missions in the 1960s and 1970s, space supporters around the world turned their sights to Mars. It took nearly seven decades to realize that dream, however, as political, financial, and technical hurdles had to be overcome to enable this next "giant leap" for our species. NASA and other international space agencies teamed up on the effort, and much like during the Apollo program, landing on the surface was achieved via a series of carefully planned intermediate steps. First, in the 2020s, robotic probes completed the global reconnaissance of the planet and then brought samples from the surface back to Earth for detailed scientific characterization and safety analysis. Then, in the early 2030s, the first

crews were sent out to travel to and orbit Mars, demonstrating the long-term capabilities of the cruise and orbiting vehicles, much like the *Apollo 8* mission had done for lunar exploration back in 1968. Then, finally, in 2037 the launch of the *Ares Ranger 3* led to the first successful human landing on Mars on July 4, 2037. Unlike the Apollo program, the *Ares Ranger 3* crew was an international mix of men and women, and included private-company employees as well as government astronauts. Billions tuned in to watch as the surface crew jumped off the landing ladder together, simultaneously planting 14 boots in the reddish, dusty soil. "*Ad astra, per viam Martis!*" they all cried in Latin—"To the stars, by way of Mars!"

A crewmember of *Ares Ranger 3* checks out the geology of Gusev crater in 2037.

An image of the *Spirit* rover, bogged down in soft, dusty soil, taken by the astronauts on the *Ares Ranger 3* crew.

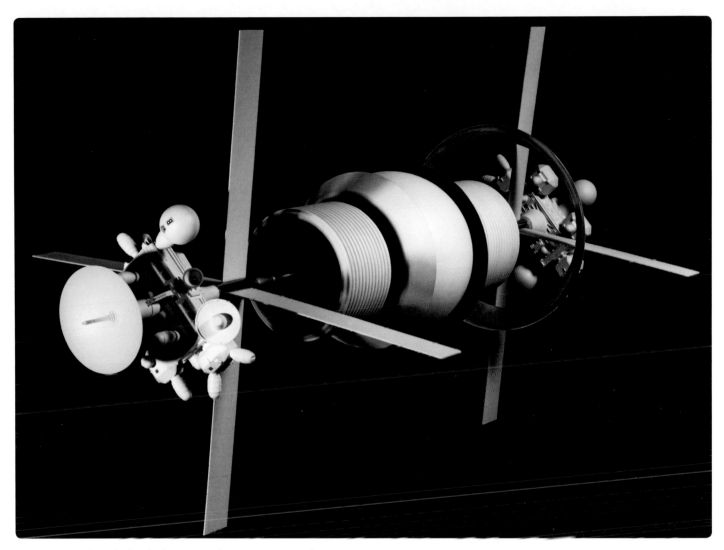

Areostation 1, launched in the late twenty-first century, is one of three (soon to be four) areosynchronous space stations orbiting high above Mars. They are used for telecommunications, scientific research, and even tourist-visit opportunities.

AREOSTATIONS 1, 2, AND 3

While they are relatively small by modern space-vehicle standards, the *Areostation* orbiting space stations—used for communications with Earth and other solar system destinations—do accommodate a small number of tourists in their facilities to help offset some of their operating costs. The small staff of each station work hard to create a friendly and fun experience that is reminiscent of what it must have been like hundreds of years ago to work and live on a relatively small space station orbiting Earth, such as the *International Space Station*. Accommodations are relatively spartan (bunks, sleeping bags), and activities focus on the nostalgic parts of space-station history—zero-gravity exercises on restored historical equipment, Mars-watching and photography from the station's large portal windows, and dining on simple, vacuum-sealed food packs that only require heating. For trained or experienced visitors, there are even occasional opportunities for spacewalks with crew members to watch them repair various components on the outside of the stations. If it's an old-school space-station experience you're seeking, book a stay on one of the *Areostations*. But beware: the wait list is long, so plan ahead.

HISTORY OF EXPLORING MARS

- **Prehistoric:** Mars recognized as a special "wandering star" in the night sky

- **1659:** Dutch astronomer Christiaan Huygens draws first telescopic maps of Mars

- **1877:** Italian astronomer Giovanni Schiaparelli draws maps of Mars showing linear features he labels *canali* (Italian for "channels")

- **1894–1909:** American businessman and astronomer Percival Lowell draws thousands of "canals" on his Mars maps; popularizes Mars as the abode of an advanced civilization

- **1965:** *Mariner 4* probe makes first flyby of Mars; reveals ancient cratered landscape

- **1971:** *Mariner 9* becomes first Mars orbiter; reveals spectacular canyons and volcanoes

- **1976:** *Viking 1* and *2* become first successful Mars landers, revealing details of the local geology and conducting first tests for life on the Red Planet

- **1997:** The *Mars Pathfinder* lander's *Sojourner* rover becomes the first rover on Mars

- **2004–20:** The *Spirit*, *Opportunity*, and *Curiosity* rovers explore Mars in detail

- **2021–25:** The *Mars 2020* rover collects and caches samples for future return to Earth

- **2028:** *Argo* robotic mission returns first samples from Mars for detailed study

- **2033:** First human crew travel to Mars orbit in the *Ares Ranger 1* mission

- **2037:** *Ares Ranger 3* crew make first human landing on Mars, in Gusev crater

- **2065:** International *Areostation 1* established in Mars orbit for research and exploration

- **2085:** First child born on Mars

- **2088:** Ares Station established as first Mars colony, in northeastern Arabia Terra

- **2110:** First routine spaceline flights begin to ferry researchers and tourists to Mars

- **2120:** Early robotic and human landing sites declared UNESCO Interplanetary Scientific Heritage Sites

- **2121:** North polar ice mining commences; "canals" built to distribute water

- **2130–70:** Three more colonies established, in Meridiani, Tharsis, and Elysium

- **2218:** United Mars Colonies recognized by the United Nations as an off-planet country

MARS ROVERS AND LANDERS INTERNATIONAL HISTORIC PARK

Much like on the Moon, historians and fans of space history have successfully called for and achieved the preservation of the earliest robotic-spacecraft and human-mission landing sites as a multinode historic park, as well as officially designated UNESCO Interplanetary Scientific Heritage Site. That's great news for tourists who are fans of space history, because it means that they can easily sign up for guided tours to visit the historic landing sites of the *Viking* landers (1976), the *Mars Pathfinder* lander and *Sojourner* rover (1997), the *Spirit* and *Opportunity* rovers (2004), the *Curiosity* rover (2012), and the *Argo* series rovers (starting in 2028). Guided trips to the landing sites of the first five human-crewed missions to Mars are also available, starting with the *Ares Ranger 3* mission, which will let you see with your own eyes the places that you've no doubt seen replayed in films and shows since you were a child. The first human landing was in Gusev crater, not far from the final resting place of the *Spirit* rover, so visiting there will get you a great two-for-one tour opportunity.

GETTING THERE

Mars and Earth have been engaged in a complex celestial dance around the Sun for billions of years, passing relatively close to each other on the same side of the Sun about every 26 months. Spaceline companies that offer the fastest transits to Mars time their departures to take advantage of these close planetary passes, offering transit times of about 6 months using traditional shuttle technologies, or as little as 2 weeks using the latest propulsion systems (at a much higher cost per passenger, of course). Typically, spacelines will arrive in Mars orbit and dock with the small moon Phobos (or less frequently, the other small moon, Deimos), and from there, a separate shuttle will take you down to one of the surface colonies or to one of the *Areostations* in synchronous orbit above the colonies.

Once you're on Mars, local tour companies offer a variety of traditional jet-powered aircraft or rocket-powered speeders to take people from colony to colony, as well as out to various tourist destinations. Budget travelers will want to steer toward the slower "areojet" services, while those who want to splurge can ride in faster rocket-powered luxury. Be sure to arrange your schedule well in advance, as bookings fill quickly and last-minute fares are, literally, astronomically priced.

THINGS TO DO, PLACES TO STAY

Mars is second only to the Moon in terms of sheer number and variety of choices for short-term and long-term visitors (as well as for permanent residents). Fans of space history; aficionados of interplanetary art, culture, and music; foodies seeking the most interesting local cuisines and creative chefs; lovers of the outdoors; couples; families—all can find great accommodations,

dining, entertainment, and recreation options on the Red Planet and its environs.

The surface bases of the four major colonies offer a wide range of accommodation and dining options, and if you're planning to spend several months on Mars (and you should—you've come all this way!), you should try to visit all four. Do some advance research, and check in with each colony's tourist representatives upon arrival to find out about any last-minute changes or deals. You can branch out from each colony to explore wonderful historical parks for hiking, camping, and photography. And don't forget to look up—visit at least one of the *Areostations* in orbit, and spend some time exploring the small moons Phobos or Deimos (or both) on your way down or up from the surface.

LOCAL FLAIR

With more than 10,000 permanent residents now living on Mars, more than half of whom were born there, a vibrant, living experiment in the future of one branch of humanity is well under way on the Red Planet. Like those who are born and live on the Moon, the lower gravity results in generally thinner, taller body shapes, lower bone density, and less muscle mass for native Martians compared to their Earthling cousins. While the rate of birth defects was high in the early days of colonization due to high surface radiation levels, great advances in technology, medicine, and habitat development have lowered that rate considerably. Ultimately, you'll notice that the locals move with a grace and style that does, indeed, seem to make them more at home there than you are.

That physical grace extends to their psychology as well, as most Martians are extremely friendly, empathetic, and welcoming when you get to know them. Perhaps their continual need to battle the natural forces aligned against them has made them genuinely want to get along with others. In the struggle for survival that is simply a day-to-day reality for humans living outside our evolutionary home planet, why waste time and energy fighting with one another?

PHOBOS & DEIMOS

TAKE A SPACE-AGE CRUISE ABOARD THE MOONS OF MARS

5

FIELD TRIP TO PHOBOS

If you're traveling to Mars, consider checking out some of the other nearby sites, such as the tiny Martian moons, Phobos and Deimos. Phobos, the innermost moon, is a frequent transfer point for passengers taking spaceline flights to Mars, and many people plan an extended layover to explore the sights on this small, lumpy world.

And it is lumpy indeed! Phobos and Deimos have been called the "potato moons of Mars" because of their generally grayish colors and spud-like shapes. Phobos is only about 14 miles (22 kilometers) across, and it has a surface area about half the size of the Samoan Islands. While there's not a lot of surface to explore, there is still a great selection of unique places to see and experience, even during a short visit.

Opposite: A popular travel poster from the early days of the Mars Colonization and Tourism Association highlights the magnificent Mars-filled view of the sky above Phobos.

Phobos is one of the stranger moons in the solar system because it orbits closer to Mars than any other known moon does to its primary planet. Phobos orbits Mars so closely, in fact, that with an orbital period of just over 7.5 hours, it actually spins around Mars faster than Mars itself rotates on its axis (which takes 24.7 hours). People on Earth see the Sun, Moon, stars, and planets rise in the east and set in the west. But people living on the surface of Mars see Phobos rise in the *west* and set in the *east* because of its rapid orbital motion. This fast pace of the moon's motion also means that people on Phobos get to see many beautiful sights on Mars move past rather quickly in the sky.

Phobos will provide you a place to get reacquainted with gravity, if you've taken a zero-*g* transport from Earth. But you will only *barely* get reacquainted, because Phobos is so small that the gravity there is 1,700 times less than on Earth! Things (including you) can fall down there—just quite slowly.

Planetary scientists are still trying to figure out where Phobos came from. Is it an asteroid that Mars somehow captured from the nearby asteroid belt? Or is it a piece of Mars itself that was flung off by a giant impact long ago? In astronomical terms, they haven't got much time to figure it out; with Phobos's orbit so close to Mars, astronomers deem it unstable and expect it to crash into the planet in about 10 million years. So, enjoy your visit to this fleeting world as you envision *that* day on the Mars of the far future . . .

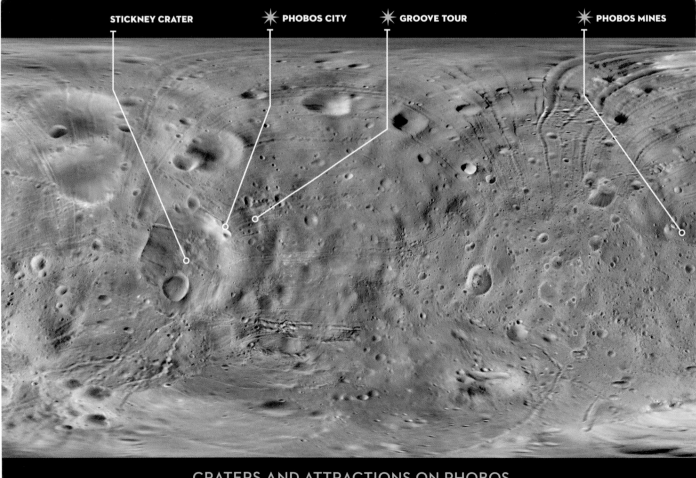

STICKNEY CRATER · PHOBOS CITY · GROOVE TOUR · PHOBOS MINES

CRATERS AND ATTRACTIONS ON PHOBOS

PHOBOS FAST FACTS

Type of Body
Planetary satellite (moon)

✳

Distance from Sun
Averages ~1.5 AU, or about 140 million mi. (225 million km)

✳

Distance from Earth
Varies over about 2 years from about 46 million–230 million mi. (74 million–370 million km)

Travel Time from Earth
Minimum of ~3 weeks, maximum of ~9 months

✳

Diameter
Irregular shape, ~17 x 14 x 12 mi. (~27 x 23 x 19 km)

✳

Highlight
The enormous impact crater called Stickney dominates one whole hemisphere.

AVERAGE TEMPERATURES

DAYTIME HIGH		NIGHTTIME LOW/SHADOWS	
°F	°C	°F	°C
25	-4	-170	-112

BEFORE YOU GO

If you're planning an extended layover or longer stay on Phobos, there are a few things you should prepare for:

Low gravity: As mentioned, while gravity exists on Phobos, it's negligible compared to its effects on Earth. You'll want to invest in a low-grav suit, including some high-quality mass boots. Aim to increase your mass by at least 10 times your Earthbound weight so that you'll have a fighting chance at semi-normal walking (more like jumping) on this low-gravity world. Alternatively, give in to the microgravity environment! Sticking with tried-and-true handrails, Velcro booties, and other zero-g tricks will get you from place to place just fine indoors on Phobos.

DON'T MISS . . .

GROOVY RIM TOUR OF STICKNEY CRATER

Phobos is covered in impact craters, but one of them, named Stickney, dominates them all, representing a huge chunk taken out of the moon's surface. Enormous grooves radiate outward from the crater, but their origin is still unknown. Regardless, at least one of the geology tour companies that operate on Phobos has decided to feature the grooves in a special new tour. Phobos/Geo Inc. signs up groups of about 20 people for the "Groove Tour," a rocking and rollicking shuttle ride up one of the main

Sadly, every year an alarming number of space tourists are injured or killed in microgravity accidents on trips to places like Phobos. Microgravity can be deceptively alluring, especially for those who don't live or work in such an environment. The thrill of being able to (literally) leap over skyscrapers in a single bound can make people feel like Superman—until they realize in horror that there's no turning back once they've made the leap. On a small world like Phobos, the gravity is so weak that it's pretty easy to set yourself into orbit, or even to escape the tiny gravity field entirely just by tripping over a stone. While modern space suits are equipped with tracking devices that will enable the authorities to *eventually* rescue you, many people who have accidentally launched themselves off a small world have died of heart attacks or space-panic syndrome long before the rescuers arrived. So please pay attention during safety drills and equipment checkouts, stay tethered during outside excursions, and show appropriate respect for the power of microgravity.

Spectacular views of Mars await from your spaceliner above Phobos.

grooves, where a local geologist provides details about the history of Phobos and its landscape, and a local DJ punctuates the science narrative with a variety of "groove" music hits. It's a quick tour, only about 90 minutes out and back, but drinks and snacks are provided to help you and your fellow travelers truly get "in the groove."

CUCINA PANORAMA

If you've only got a short layover, or just aren't keen on wandering too far from the spaceport, a highly recommended way to while away a few hours is to book a meal at the Cucina Panorama, one of the oldest restaurants in the Mars system. With seating under a massive transparent dome, you and your fellow diners can watch as the surface of Mars—which nearly fills the sky as it hangs above you—drifts elegantly past. The vibe is reminiscent of one of Earth's old rotating skyscraper restaurants, except the whole world is literally sweeping past above your head. The food is so-so (dominated by synthetics and other quick eats preferred by travelers on the go), but that's not why you're there. The view is all, and even though the price tag for a table will be steep, it's a spectacular experience.

WHY IS PHOBOS SO GROOVY?

American astronomer Asaph Hall discovered Phobos in 1877, and when Phobos was finally photographed up close by space probes in the twentieth century, its largest geologic feature was named after Hall's wife, Angeline Stickney Hall. Stickney is interesting to planetary geologists because the impact event that created it must have almost ripped poor little Phobos completely apart. But it didn't, and the resulting crater dug deep into the subsurface, creating a huge hole in the ground. Strange grooves radiate out from Stickney, and beautiful bright and dark patterns of erosion decorate the crater's interior walls. The grooves, though, are what mystify geologists, partly because they stretch more than halfway around the surface of the entire moon. Could they represent tectonic faults and valleys where Phobos started to be ripped apart under the stresses from the impact that created Stickney? Or are they long gouges created by boulders and other debris that flew out of the hole and carved up the ground as the crater was being formed? Geologic surveys to collect and analyze samples still haven't settled the issue.

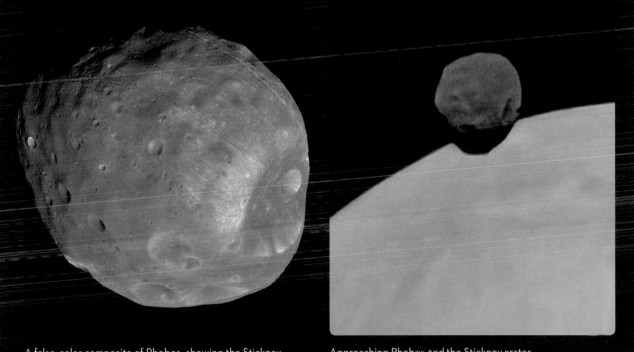

A false-color composite of Phobos, showing the Stickney crater (the big divot on the right).

Approaching Phobos and the Stickney crater.

PHOBOS MINES INC.

It's been known since the early space age that Phobos is rich in so-called carbonaceous materials (like certain classes of meteorites). The discovery that large deposits of carbon-rich ore and significant amounts of water bound up in those minerals could be mined from some of the far-side craters provided the driver for what has emerged as the major industry on Phobos. The locals (known as Phobians) consume much of the water on-moon but send most of the carbon all over the solar system,

HISTORY OF EXPLORING PHOBOS

1877: Phobos discovered by American astronomer Asaph Hall

1971: First low-resolution images of Phobos taken from space by the *Mariner 9* probe

1977: First high-resolution images of Phobos taken by the *Viking 1* Mars orbiter

1989–2010: Additional high-resolution images and mapping conducted by the *Phobos 2, Mars Global Surveyor, Mars Express,* and *Mars Reconnaissance Orbiter* spacecraft

2005: First image of Phobos from the surface of Mars, taken by the *Spirit* Mars rover

2029: First successful robotic sample-return mission from Phobos (*Phobos 3*)

2035: First astronauts land on Phobos, as part of the *Ares Ranger 1* Mars orbiter mission

2078: First successful carbon and water extraction demonstrated on Phobos

2110: Phobos City founded as the primary spaceport on Phobos

2112: Phobos Mines Inc. starts industrial-scale carbon and water mining on Phobos

2218: Stickney Rim Trail and Park becomes part of the Interplanetary Park System

where it is used as raw material for manufacturing (steel, carbon-fiber nanotubes, and other advanced construction materials), electronics, synthetic diamonds, and even the now-famous sleek graphite pencils that have become iconic of Phobos Mines.

GETTING THERE

The duration of your trip to Phobos (and Mars) will depend on when you want to leave and how much you can afford to spend. If Mars and Earth are on the same side of the Sun and you can spring for seats on one of the new, pricey hyperdrive nuclear-powered shuttles, you can get from Earth to Mars in as little as three weeks. On the slow, economy-priced end, if you depart during the traditional once-every-26-months launch windows like those used for the first missions back in the twentieth century, a chemical propulsion shuttle will take anywhere from 6 to 9 months to get you there. Most of the major spacelines will stop at either Phobos or Deimos (or both) on their way to or from Mars. Be aware, however, that

the frequency of Mars nonstops has been increasing in the past few years. Regardless of the carrier, you'll arrive in Phobos City, the largest (and, really, only) major settlement and spaceport, located along part of the rim of Stickney crater on the Mars-facing side of the satellite.

THINGS TO DO, PLACES TO STAY

Phobos is mostly an industrial town and transfer point, so there aren't a lot of tourist activities or resort-style accommodations to choose from. There is a small but lovely visitor's center and museum near the carbon and water mines on the far side of Phobos, but getting there can be a challenge, so plan ahead. Luckily, though, you don't have to stray outside the underground walkways and plazas of Phobos City to find decent places to eat, drink, and stay (mostly 2- and 3-star hotels, all with standard low-g pools, many with casino games, and even a few with terrariums and gardens). Dining and drinking options can be a shot in the dark, unfortunately, as restaurant and bar owners often get lured away to the looming Red

Planet below to try to make their fortunes. Despite that, a few staples have remained over the decades, including Phobos Pharms, which advertises "the best vegetarian food in the entire Martian gravity well" (they claim this is a by-product of the carbon-rich soil's ability to hold nutrients so well), and the watering hole known as Fear Not, which is said to have served drinks to the first twenty-second-century colonists. Check with the information desk at the disembarking terminal when you arrive to learn about the latest establishments and reviews.

LOCAL FLAIR

Phobians are extremely proud of the safety and productivity records of the carbon and water mines on Phobos, so if you have friends there—or make some during your visit—you just *might* be able to arrange for a rare VIP tour of the facilities. It's worth trying to get a sneak peek inside the inky black shafts and conduits of the inner solar system's largest and most efficient carbon mine. Yet another reason to strike up a conversation with the people you meet, who can teach you more about the local customs and culture.

Early astronauts took their chance on low-g Phobos using only weighted anchor boots. Be more safety-conscious on your trip, and rely on tethers and the experience of your guides.

SMOOTH JAZZ ON DEIMOS

Lots of visitors to Phobos also plan outings to the other small, asteroid-like moon of Mars, Deimos. While Phobos is considered by most to be *the* gateway to Mars, slightly smaller, more distant, and smoother Deimos is becoming a popular resort destination of its own, especially among culture enthusiasts and interplanetary musicians. Indeed, for decades the Deimos Tourism Board has been using the smooth nature of the moon's surface and its slow circuit around the Red Planet to specifically attract the slow and suave sounds of smooth-jazz musicians. The Deimos City Jazz Festival, for example, attracts the solar system's best musicians each Mars year.

Opposite: Part of the terrarium and habitat section of the Deimos Ring
Station as it orbits above Deimos City.

While Phobos orbits close to Mars and spins frenetically around the planet in only about 7.5 hours, Deimos orbits more than twice as far away and takes over 30 hours to travel once around the Red Planet. Indeed, because the orbital period of Deimos is only a few hours longer than the spin period of Mars on its own axis, the planet appears to hang almost motionless in the Deimos sky. Similarly, observers on Mars see Deimos move very slowly across the Martian sky, taking almost 3 sols (a day on Mars, 24 hours and 37 minutes long, is called a *sol*) to rise in the east and then set in the west. Deimos's slow-motion dance relative to Mars has translated directly into a destination where the attitude is similarly slow, mellow, and relaxing. "Slow down—you're on Deimos" is a common greeting for newly arriving visitors.

The other big difference between Phobos and Deimos is Deimos's topography and surface texture. As discovered back in the early space age, Deimos is much smoother than Phobos. While it has an ancient cratered surface, most of those craters, as well as the surrounding plains, are covered by smooth, almost powdery soil. Planetary scientists call that soil *regolith* when it appears on small airless bodies, and it's abundant on Deimos. Some process has created a large supply of this finely ground rock and soil and distributed it across the surface, muting the topography of some craters and completely filling in others. Geologists are still trying to figure out how that happened on Deimos and why it didn't happen on Phobos. While Deimos was the twin brother of Phobos in ancient Greek mythology, in reality they are truly quite different little worlds.

BEFORE YOU GO

If you're planning a visit to Deimos, there are a few things you should prepare for:

Low gravity: Just like on Phobos, the pull of gravity on Deimos is tiny compared to its effects on Earth. See the discussion in chapter 5 to get ready for this low-*g* world.

Soft soil: Finely ground, almost powdery soil covers much of the surface of Deimos, and in some places the powder is many meters deep. The first astronauts to explore the surface of Deimos learned that equipment and even people can slowly sink into that fine powder if not properly anchored or tethered. Because of its flour-like consistency, the soft soil can also easily gum up motors, gears, and airlock seals, much like the famous and similarly fine reddish dust on nearby Mars. To avoid such potential dangers, your tour guides will keep you and your support equipment on well-marked (and well-tamped) trails if you're out and about on the surface. Do not stray!

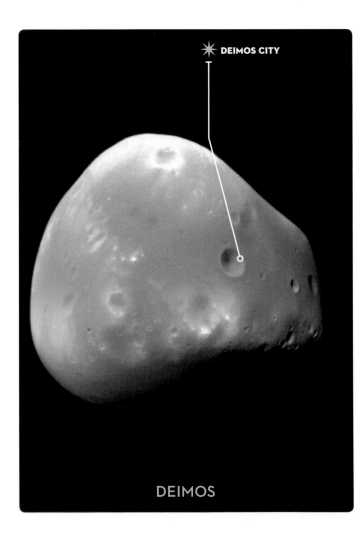

DEIMOS CITY

DEIMOS

DEIMOS FAST FACTS

Type of Body
Planetary satellite (moon)

✳

Distance from Sun
Averages ~1.5 AU, or about 140
million mi. (225 million km)

✳

Distance from Earth
Varies over about 2 years from
about 46 million–230 million mi.
(74 million–370 million km)

Travel Time from Earth
Minimum of ~3 weeks, maximum of
~9 months (most tourists arrive via a
shuttle ride from Phobos, however,
which takes 1–6 hours)

✳

Diameter
Irregular shape, ~9 x 7 x 7 mi.
(~14 x 12 x 11 km)

✳

Highlight
The surface is smooth with features
blanketed or buried by fine-grained
rocky soil.

AVERAGE TEMPERATURES

DAYTIME HIGH		NIGHTTIME LOW/SHADOWS	
°F	°C	°F	°C
25	-4	-170	-112

DON'T MISS . . .

DEIMOS RING STATION

🍴 📷

Deimos Ring Station—or just *the Ring* as the locals call it—is a spinning circular space station, 1,000 feet (300 meters) wide, that is in orbit just a few miles above Deimos City (though it is considered part of the city). The Ring, built in the mid-twenty-second century, spins once every 30 seconds to create an Earth-gravity environment. Resort hotel rooms, restaurants, music clubs, a spa and terrarium, and numerous museums and concert performance spaces line the 0.6-mile-long (1-kilometer) inner track. The Ring can house about 2,500 visitors

and staff at maximum capacity, which it usually reaches during and around the jazz festival. Venues on the Ring offer spectacular views of Mars, Deimos, and occasionally Phobos, and high-end Ring restaurants routinely compete for best cuisine in the Mars system.

DEIMOS CITY JAZZ FESTIVAL

👪 🍴

For a week on each side of the Martian New Year holiday (which comes about once every 26 Earth months), little Deimos becomes the undisputed interplanetary jazz capital of the solar system. The festival originated in the mid-twenty-second century as a small gathering of musicians in residence on the newly built Deimos Ring

WHERE DID DEIMOS COME FROM? (OR PHOBOS, FOR THAT MATTER.)

Planetary scientists have two major hypotheses to explain the existence of the two small, lumpy worlds around the Red Planet. One idea is that they are asteroids that were captured into orbit by Mars. The gravity of Jupiter or other planets could have nudged some small, lumpy bits of rock, metal, or ice out of the Main Belt of asteroids and into orbits that crossed the orbit of Mars, which is close to the inner edge of the Main Belt. Some would have crashed into Mars, others would have been diverted out of the solar system entirely by near misses with Mars, but a very small and lucky couple could have been captured by Mars—stolen, if you will, from the

Main Belt. The competing idea is that both Phobos and Deimos are pieces of Mars itself, formed from the re-accretion of debris jetted off Mars when a large asteroid or comet crashed into the planet long ago. In that second scenario, Mars would likely have had a ring of rocky material from which the small moons would have formed over time. The last bits of fine-grained ring material slowly getting swept up by Deimos could then explain the prevalence of smooth, powdery surfaces there. The arguments continue, however, as neither hypothesis has yet been able to explain everything we know about both Phobos and Deimos.

Racing PowderShips—super-fast, single-rider rockets designed for the smooth, powdery surface of Deimos—is an exciting activity for visitors to the smallest moon of Mars.

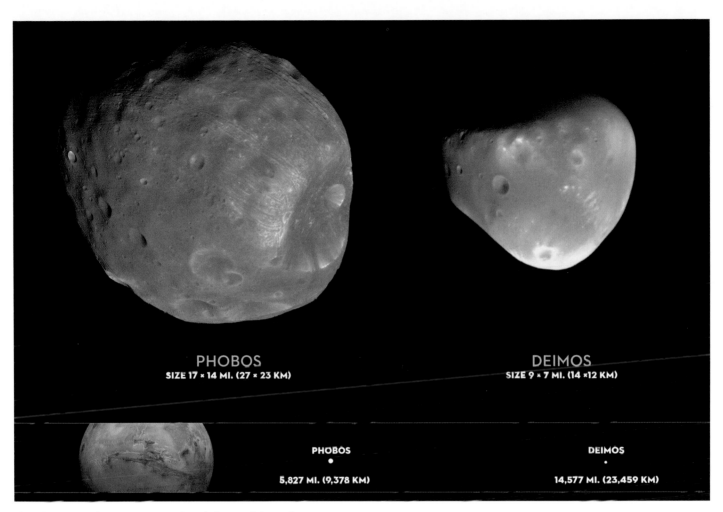

PHOBOS
SIZE 17 × 14 MI. (27 × 23 KM)

DEIMOS
SIZE 9 × 7 MI. (14 ×12 KM)

PHOBOS
5,827 MI. (9,378 KM)

DEIMOS
14,577 MI. (23,459 KM)

The relative size of Deimos compared to Phobos, and their relative distances from Mars.

Station. Word got out about the new experimental twists the locals were applying to the centuries-old art form, and the organizers started to attract famous musicians from Earth, Mars, and other colonies who wanted to be part of the jazz renaissance. That renaissance continues today, not only with live performances on the Ring but also with impressive low-*g* jazz concerts on Deimos itself. Smaller festivals and performances are streamed across the solar system year-round.

POWDERSHIP TOURS AND RACES

PowderShip touring and racing is a relatively new way to turn the hazards of Deimos's dust into a relatively fun and generally safe (for the public) recreational activity.

PowderShips skim across—and sometimes *within*—the loose powdery soil of Deimos at high speed, like Florida fan boats skimming through the high swampy reeds back on Earth. It's a crazy rush as you and your fellow passengers power through the powder at high speed, catching glimpses of Mars looming high above now and then when the dust momentarily clears. All of the public tour courses have been thoroughly scanned for rocks or other obstacles, but the professional raceways have both natural and intentional obstacles that test the nerves and reaction times of even the best drivers. Indeed, several recent controversial crashes and fatalities have many calling for new safety standards and regulations to keep racers from getting hurt.

HISTORY OF EXPLORING DEIMOS

- **1877:** Deimos discovered by American astronomer Asaph Hall
- **1972:** First low-resolution images of Deimos taken from space by the *Mariner 9* probe
- **1977:** First high-resolution images of Deimos taken by the *Viking 1* Mars orbiter
- **2004:** First studies of Deimos solar transits from the surface of Mars, from the *Spirit* rover
- **2009:** Additional high-resolution images and mapping by the *Mars Express* and *Mars Reconnaissance Orbiter* spacecraft
- **2035:** First successful robotic sample-return mission from Deimos (*Deimos 1*)

- **2047:** First astronauts land on Deimos as part of the *Ares Ranger* 6 Mars mission
- **2078:** Global radar mapping identifies depth of the soft soil across Deimos
- **2110:** Deimos City founded as the primary spaceport on the surface of Deimos
- **2152:** Deimos Ring Station completed, orbiting above (and considered part of) Deimos City
- **2160:** Deimos City Jazz Festival established, primary venue on Deimos Ring Station
- **2218:** First PowderShip rides established, in plains surrounding Deimos City

GETTING THERE

Very few spacelines offer direct flights to Deimos City from Earth, because most travelers use the more well-developed Phobos City spaceport as a stopping point on their eventual trip down to the surface of Mars. Nonetheless, there are frequent shuttle flights between Phobos and Deimos, typically taking only one to six hours depending on the relative distance between the two moons and the kind of shuttle you book. If you time it right, you can even make your visit to Deimos a quick day trip, catching an afternoon concert in Deimos City and getting back to Phobos in time for dinner. A longer stay to take in more of the sights (and sounds) is highly recommended, however.

THINGS TO DO, PLACES TO STAY

Deimos City is perhaps the inner solar system's finest off-Earth artist colony, with painters, writers, sculptors, and especially musicians flocking to the sparsely inhabited moon to ply their arts in a less hectic, less industrial, and, yes, less touristy environment compared to Phobos. Locals have constructed special museum and performance venues on and above the surface specifically designed to accommodate visitors.

Standard 2- and 3-star hotel rooms and condo-style apartments can be rented in Deimos City for low-*g* accommodations. There are a few 5-star resorts and restaurants on the Ring, and most of the music and art shows, including the Deimos City Jazz Festival, are held there. In addition to a wonderful but relatively "normal" Earth-gravity music experience on the Ring, try to take in a low-*g* concert on the surface of Deimos itself. It's a very different vibe, with floating musicians, specially made versions of classical instruments that were originally built for Earth gravity (like a baby grand piano), and lots of deep reverberation off the thick habitat windows making for a unique audio and visual experience. Frequent shuttles ferry passengers to and from the surface and the Ring.

Only a few industries have popped up on Deimos so far, mostly focused on musical instrument sales and repair, extraction and processing of the fine-grained soil for building materials, and some mining of low-grade carbon and water for use by the locals.

LOCAL FLAIR

The few thousand permanent residents of Deimos (known as Deimosians) are a split between artists/performers and the service industry workers who help manage visits and events involving the general public. Befriending either can sometimes pay off via dinner-party invitations with some of the local artists, deals on great seats at shows (most of the locals are season-ticket holders), or even help getting any last-minute tickets at all. Good luck getting the locals to give up their own coveted dinner-table seats during the Jazz Festival on the Ring, however.

Mars looms large and nearly motionless in the skies above smooth, powdery Deimos.

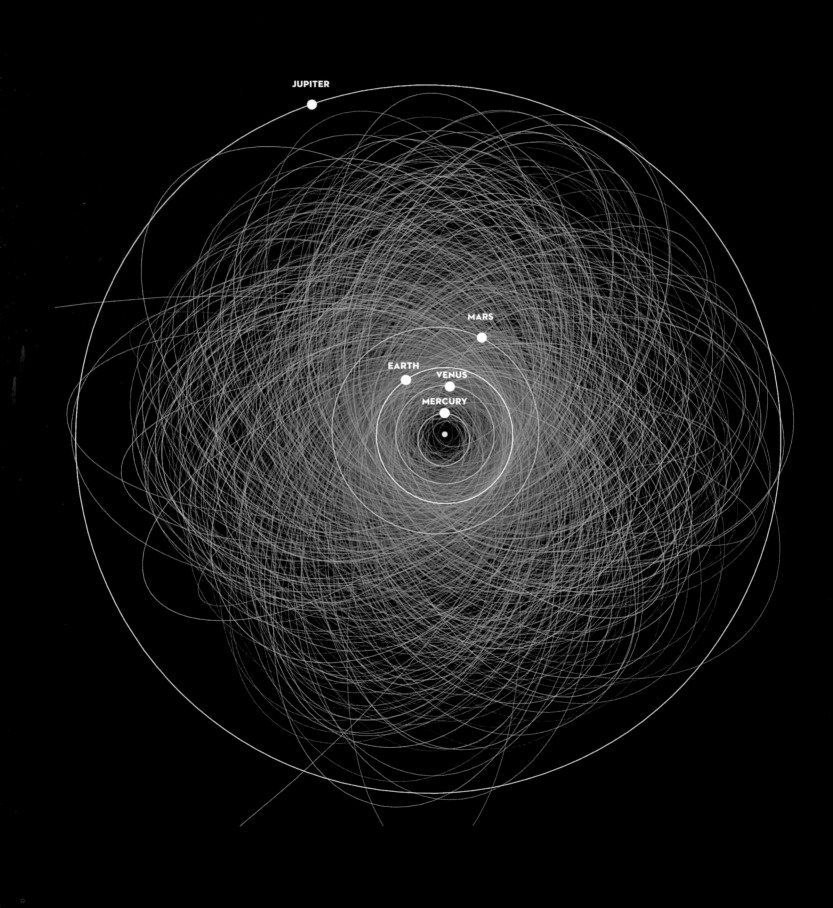

CLOSE ENCOUNTERS WITH NEAR-EARTH ASTEROIDS

We tend to think of our solar system as just our Sun and a small collection of planets and their moons. Thus, it's easy to forget that there are literally millions of other small worlds out there that wander among, and sometimes into, the planets: asteroids. The word *asteroid* means "star like," because when the first asteroids were discovered in the early nineteenth century, they looked like mere points of light in astronomers' telescopes. Since the late twentieth century, however, we've been able to visit many of these little worlds on space missions and learn about their surfaces, interiors, and origins. And now we can visit many of them ourselves!

Opposite: Between the orbits of Mercury (inner circle) and Jupiter (outer circle) is a swarm of Near-Earth Asteroids (NEAs, blue points in diagram) that travel near—and sometimes impact—Earth and other inner planets.

The *NEAR Shoemaker* spacecraft wasn't designed as a lander but was nonetheless gently guided to a descent on the surface of Eros way back in 2001. Once there, it transmitted data for a few days.

One particularly interesting and fun class of asteroids is called the Near-Earth Asteroid (NEA) population. These are small, rocky bodies, ranging in size from a few hundred to a few thousand feet across, that orbit the Sun in paths that occasionally bring them close to Earth. Some of them even cross our planet's orbit during their trip around the Sun, making them what astronomers call Potentially Hazardous Objects, because they could eventually crash into our home world. Impacts by the smallest NEAs produce spectacular fireballs but no major damage; impacts by the largest NEAs could significantly alter life as we know it. Just ask the dinosaurs . . .

While the potential threat of NEAs has made them important targets for scientific study, their proximity has made them compelling tourist destinations for the modern solar system vacationer. Three asteroids in particular—Eros, Itokawa, and Apophis—have become popular destinations because of their occasional very close passes near Earth, and thus the short travel times to get to them. Some tour companies have taken a "wait and see" approach: wait for a new, small NEA that's about to pass close by Earth or the Moon, and then arrange a charter spaceline flight to go out and see it!

NEAR-EARTH ASTEROIDS FAST FACTS

Type of Body
Small, rocky asteroids

✳

Distance from Sun
When near Earth, ~1.0 AU, or about
93 million mi. (150 million km)

✳

Distance from Earth
Closest encounters can bring them
closer than the moon, and some have
impacted our planet

Travel Time from Earth
A few hours to days for ones that
pass super close; a few weeks to
months for more distant encounters

✳

Diameter:
Irregular shape, ranging from a few
kilometers across to a few tens of
meters across

✳

Highlight
Help tag newly discovered NEAs so
scientists can monitor their orbits!

AVERAGE TEMPERATURES
(USING EROS AS A TYPICAL EXAMPLE)

DAYTIME HIGH		NIGHTTIME LOW/SHADOWS	
°F	°C	°F	°C
212	100	-238	-150

BEFORE YOU GO

If you're planning to visit one or more of the many
Near-Earth Asteroids out there, there are few things you
should prepare for:

Low Gravity: Near-Earth Asteroids are very
small worlds, and so, like Phobos and Deimos, they have
very little gravity. Some of the super-small ones that you
can visit—lumps of rock only as large as a football field—
have no significant gravitational pull. So don't expect to
"walk" on these worlds. Encountering them is more like
docking with a space station than landing on a planet,
so you'll need to be equipped with a good space suit. If
you're lucky, previous explorers or your tour guides will
have installed pitons on the asteroid so that you and a
group of others can climb or rappel across the rocky

surface without worrying about floating off into space.
Even though you can't get hurt from falling during this
kind of rock climbing, be sure to verify the safety of your
equipment, lest you drift off and have to be rescued.

High Spin Rate: While some NEAs spin slowly
over the course of many hours, like most moons and
planets, others are known as rapid rotators because they
take minutes, not hours, to spin once on their axis. Visitors
should approach these objects with extreme caution,
because most of them are solid boulders, often with sharp
edges or even glassy compositions. Trying to grab onto
such a fast-spinning hunk of mass can cut your space suit,
break a bone, or fling you at high speed out into space.
Follow your tour guide carefully!

ASTEROIDS AND METEORITES ARE CONNECTED

Asteroids crash into Earth all the time—it's just that most of them are really small, ranging in size from boulders to flecks of dust. Most of them burn up in Earth's atmosphere, creating fiery streaks across the sky that we call shooting stars. Pieces of some of the larger ones can make it down to the surface intact, though. When scientists find some of those pieces, we call them meteorites. In general, we don't know where specific meteorites came from. Are they chunks of larger asteroids that were knocked off in impact events on other so-called parent bodies? Are they primitive materials that formed right out of the cloud of gas and dust that formed the rest of the planets and moons in our solar system? Fortunately, we can identify the origins of some of them by comparing results from laboratory studies of meteorites with telescope and spacecraft studies of asteroids and making direct connections between some members of those populations. Asteroids such as Eros, for example, which are called S types because of their silicate composition, appear to be the parent bodies of the most common kind of meteorite class, ordinary chondrites. Some meteorites in our collections are from the Main Belt asteroid Vesta; others are from the Moon or Mars.

DON'T MISS . . .

EROS

📷 📖 🍴

Eros was discovered back in 1898 and is among the most famous of the NEAs because of its occasional close passes by the Earth and its status as the first asteroid to be studied up close by a space mission. Between 1999 and 2001, NASA's Near-Earth Asteroid Rendezvous (NEAR) mission flew by, orbited, and eventually landed on Eros. The *NEAR Shoemaker* spacecraft is still there on Eros and is one of the most popular destinations for NEA tourists. The spacecraft and landing site are designated Interplanetary Historic Landmarks, so if you wish to visit, you must take one of the organized tours specifically approved to (carefully) hover over the site. Photographers and space historians are continually impressed at how pristine the spacecraft and its gleaming solar panels appear after being on the surface for hundreds of years.

Your guides will also take you for a fun tour across the surface of the oblong, cigar-shaped, twice-the-size-of-Manhattan asteroid, letting you see up close the craters, ridges, and dust-filled depressions that were mapped out by planetary geologists back in the day. Plan to spend at least a few days on *Eros 1*, the spinning space station that tags along with Eros and provides Earth-gravity lodging, dining, and entertainment options for visitors to this small world. Perhaps because it is named after the Greek god of love, Eros has become an extremely popular destination for honeymooners, and *Eros 1* station offers ample opportunity for romantic dinners, dancing, shows, and even a zero-g spa that caters to couples.

Left: Eros, the first NEA discovered. It is 21 miles (34 kilometers) long, about the size of Lake Geneva.

APOPHIS

Apophis is another famous Near-Earth Asteroid, and it is aptly named after the Egyptian god of destruction, because it is one of the largest known objects that is almost certain to impact the Earth in the future. Apophis caused a small scare shortly after it was discovered back in 2004, when some astronomers thought it had a decent chance of impacting Earth in 2029 or 2036. While additional observations showed that would not be the case, Apophis does nonetheless make occasional, very close passes by Earth. It's a rocky body about three-and-a-half football fields across, traveling at more than 67,000 miles per hour (108,000 km per hour), and so it would cause significant local and regional damage wherever it strikes. Fortunately, astronomers studying Apophis have since ruled out any chance for an impact with Earth in the next few hundred years. They've been able to determine that by deploying small radio-transmitter devices, or "tags," onto the surface of Apophis to help track its orbit with great accuracy. Once or twice a decade those tags need to be replaced, providing an opportunity for ecotourists to accompany the scientific crews on their tagging expeditions. While the opportunities are rare, the price is high, and the accommodations are light-years from 5-star, it's a great opportunity to pitch in and learn about asteroids, impact hazards, and planetary science in general. So consider tagging along!

Left: The small 1,755-foot-long (535-meter) NEA Itokawa, which had samples returned to Earth by the Japanese *Hayabusa* ("Falcon") space probe in 2010, has a loose, rubbly surface perfect for mining.

Right: The largest boulders pictured on this zoomed-in image of Itokawa are about 6 feet (2 meters) across.

ITOKAWA

While a pretty small NEA, Itokawa gained global fame centuries ago, in 2010, by becoming the first asteroid from which samples were gathered and returned to Earth. The unmanned Japanese *Hayabusa* mission brought back tiny amounts of dust from the surface of this five-football-field-long world, providing rich details about the chemistry and mineralogy of small, rocky asteroids. *Hayabusa* also discovered that Itokawa is a special kind of asteroid called a rubble pile—that is, it's not a coherent piece of rock but a collection of millions of boulders, rocks, and sand grains held together by their mutual gravity. Because of that, and its proximity to Earth, in the twenty-third century Itokawa has become a major supply station for rock-based construction and building supplies, especially the materials needed to make concrete for structures on off-world destinations. There are limited opportunities for tourists to accompany the frequent cargo and equipment transports that run to and from Itokawa to bases on the Moon and elsewhere. If you visit it, in addition to learning about the geology of small worlds like Itokawa, you'll get a chance to see some of the latest technologies in asteroid-mining equipment up close.

GETTING THERE

It has been known since the early space age that it is easier to rendezvous with some Near-Earth Asteroids than it is to land on the Moon. This is because a now well-known subset of a few hundred of these small bodies routinely make very close passes through the Earth-Moon system, and so a well-timed launch from Earth (or the Moon) can, within days, easily encounter these little worlds when they happen to zip past. In 2045, the first human crew to visit an asteroid exploited this convenience, as their target, 2006 RH120, was one of those special, easy-to-reach NEAs. It has been exploited again recently with a mission to another close-passer, the metal-rich boulder known as 2008 UA202, which was just captured and returned to Earth by some daring metals-market speculators.

Longer, more traditional spaceline visits to NEAs are also possible, as a variety of tour companies offer trips to small worlds like Eros and Itokawa throughout the year. Depending on the positions of the asteroids relative to Earth, trips can take anywhere from a few weeks to a few months each way.

THINGS TO DO, PLACES TO STAY

Unlike many other solar system destinations, there are as yet very few "get out and about" options when visiting Near-Earth Asteroids, mostly because they have such low gravity that it's impossible to walk across their surfaces in a normal way. Still, if you're game to try tethering or low-*g* rock "climbing" after some training, there are many such options for actually going out and being on—or more like *with*—the surfaces of these small worlds. Your accommodation choices will also be limited, with *Eros 1*

Robotic asteroid-tagging probes like this one (based on the early-twenty-first-century *OSIRIS-REx* probe design), shown here placing a radio transmitter on the potentially hazardous NEA Apophis, have been phased out in favor of astronaut missions, which now deploy more robust, long-lifetime radio tags.

HISTORY OF EXPLORING NEAR-EARTH ASTEROIDS

- **1801:** First asteroid discovered (Ceres) in the Main Asteroid Belt between Mars and Jupiter
- **1898:** First Near-Earth Asteroid (Eros) discovered
- **1908:** Atmospheric explosion from the impact of a small NEA or comet devastates forest around Tunguska, Siberia
- **1968:** First NEA detected with Earth-based radar
- **1999–2001:** *NEAR Shoemaker* mission flies by, orbits, and lands on Eros
- **2000:** Number of known NEAs exceeds 1,000
- **2010:** Japanese *Hayabusa* mission returns first NEA samples (from Itokawa) to Earth
- **2013:** Number of known NEAs exceeds 10,000
- **2021:** First privately funded spacecraft prospecting of asteroids (for water and metals) begins
- **2023:** NASA's *OSIRIS-REx* mission, launched in 2016, returns sample from Bennu, a carbonaceous NEA

- **2030:** Number of known NEAs exceeds 1,000,000
- **2045:** First human-crewed mission to a Near-Earth Asteroid (2006 RH120)
- **2065:** Orbits and threat potential of all possibly hazardous NEAs known
- **2071:** Astronauts land on NEA Itokawa and demonstrate low-*g* drilling, mining technology
- **2085:** Astronauts land on Apophis and establish permanent tagging station
- **2104:** First tourist expedition to a Near-Earth Asteroid (Eros)
- **2160:** First water extraction plant set up on a Near-Earth Asteroid (Bennu)
- **2198:** *Eros 1* space station and base begins operations close to Eros
- **2218:** First small, metallic NEA (2008 UA202; 16 feet [5 meters] across) captured and returned to Earth

station being the closest thing to a Moon- or Mars-like base or space-station hotel. For other asteroids, you'll likely be eating, sleeping, and getting your entertainment on the spaceliner itself, so be sure to choose wisely, especially if you've signed up for a long trip out and back.

LOCAL FLAIR

The crew and employees of *Eros 1* (*Erosians*, as they call themselves) are so far the only "inhabitants" of the Near-Earth Asteroid population, but they've already come up with a variety of customs and traditions that some of them, at least, hope will spread to other future NEA bases and stations. These include a preference for cooking with water extracted from Bennu, a small and (relatively) nearby asteroid that was discovered back in the twenty-first century to consist of rocky, carbon-rich material with a small amount of water in it. Robotic mining operations now extract that water from Bennu's soil, and Bennu Mines Inc. ships it all over the solar system. A few venturesome Erosians even use Bennu water and hydroponically grown ingredients to make "locally sourced" chocolates for newlywed couples to enjoy on the station.

SUN
Come see Sol - the Center of the UNIVERSE!
Twenty Four Hour Tanning Salons!
GUARANTEED NOT TO EXPLODE
FOR ANOTHER 4,999,999,723.2 YEARS!

GET SOME SUN!

If the winter blues are getting you down, a good way to perk up is to get out of town and get some sun, right? Well, thanks to a unique new opportunity being offered by Solar System Tours Inc., you can actually get out of town and *literally* get some Sun. Specifically, the company is now booking three-week tours on a specially equipped luxury spaceliner that will take you closer to the visible surface of the Sun than any pleasure craft ever before. If you're lucky, the ship will even cruise through a spectacular magnetic loop or prominence that is more than a dozen times larger than Earth. What a sight!

The Sun is a typical star: an enormous nuclear reactor that fuses groups of four hydrogen atoms into single helium atoms deep in its core, and releases a small amount of energy with every new helium atom created. That energy makes its way out of the visible surface of the Sun (the photosphere) as the light and heat that powers the surface conditions and weather on the planets of our solar system, and that ultimately serves as the driving force for life on Earth.

Opposite: Don't forget a visit to the Sun as part of your solar system vacation plans!
(*Sun*, illustration by Indelible Ink Workshop)

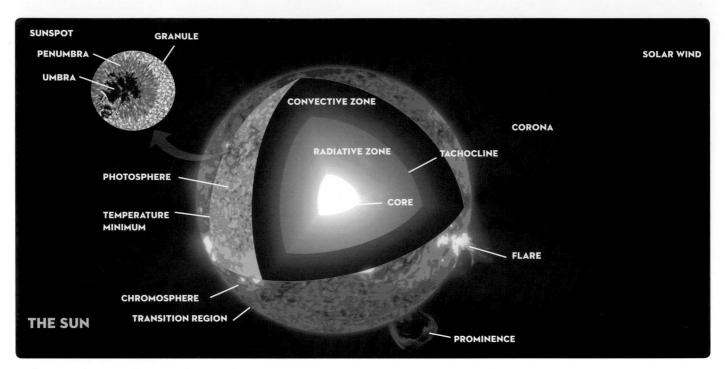

The various features and regions of our star, the Sun.

Until recently, spaceprobes couldn't get very close to the photosphere of the Sun because of its extremely high temperature, as well as the much higher temperatures of the Sun's outer atmosphere, or corona (the spectacular radial structures that can be seen during total solar eclipses). However, recent advances in spaceliner materials, thermal control systems, and radiation shielding now make it possible to safely visit such places for relatively short periods of time. Astronomers who study the Sun have taken advantage of these new technologies to probe our star in greater detail than ever before, using robotic and crewed missions. And now, adventure tourists can take advantage of these new technologies too!

BEFORE YOU GO

If you're planning to take an adventure tour close to the visible surface of the Sun, there are a few things you should prepare for:

🌡️ **High Temperatures:** Obviously, the Sun is hot. And the outer hull of your spaceliner will get extremely hot, too, as you pass through the corona and approach the photosphere. The ship's technology is designed to protect the people and systems inside, but failures do occur and backup systems can only help so much. In the event of a primary thermal system anomaly, your spaceliner crew will engage the backup systems and speed the ship away from the Sun as quickly as possible. Still, the inside of the ship will heat up to uncomfortable levels—modern thermal backup systems are typically able to cool the environment down to only about 122 degrees Fahrenheit (50 degrees Celsius) that close to the Sun. These modified spaceliners do have special ice pods for crew and passengers, however, which can serve as refuges for the hour or so that it takes to make an emergency ascent away from the Sun's hottest regions. Pay attention during the safety drills!

DON'T MISS . . .

ARTIFICIAL SOLAR ECLIPSES

📷

While total solar eclipses are rare for people on the surface of Earth, they can be routinely "created" for satellites and vehicles out in space by simply occulting

SUN FAST FACTS

Type of Body
Star

✳

Distance from Earth
Averages 1.0 AU, or about 93 million mi. (150 million km)

✳

Travel Time from Earth
Typically about a week to a week and a half using new-technology spaceliners

Diameter
~865,000 mi. (1.4 million km), with a volume that could hold about a million Earths

✳

Highlight
While a typical star (just one of about 100 billion stars in our Milky Way galaxy), the Sun is the source of all light and heat that makes life on Earth possible!

AVERAGE TEMPERATURES

	REGION OF THE SUN	
	°F	°C
Visible surface ("photosphere")	10,300	5,700
Thin outer atmosphere ("corona")	5,400,000	3,000,000
Core of the Sun	27,000,000	15,000,000

CAUTION!

The field of study of the minute-by-minute to year-by-year changes in our dynamic Sun is called space weather, because those changes affect the stream of high-energy particles and magnetic fields that emanate from the Sun and influence Earth and other planets. (The polar auroras, or northern lights, for example, are a manifestation of space weather.) You should plan for the "weather" to be unpredictable as you approach the Sun, and your pilots and crew may have to react suddenly to rapid changes in the Sun's local temperature or radiation output from sunspots, solar flares, or prominences. This could mean sudden course corrections that create high-gravity acceleration, sudden changes in the configuration of thermal-control or radiation-shielding systems, or the need to rapidly take refuge in your ice pod. Tours close to the Sun are characterized as adventure travel for just these reasons!

TOTAL SOLAR ECLIPSES

Total solar eclipses—when the Moon's disk temporarily blocks the disk of the Sun—are among the rarest astronomical phenomena that can be experienced. On average, any place on the surface of Earth will experience a total solar eclipse only about once every 300 years, and so most people have never seen one. During the precious few minutes of totality as the shadow of the Moon passes over, the very faint corona—the outer atmosphere of the Sun—can be seen and studied, because the much more intense light from the Sun's photosphere is being blocked. In fact, the element helium (from the Greek personification of the Sun, Helios) was discovered during a total solar eclipse back in 1868.

So-called "eclipse chasers"—tourists who seek out the thrill and photogenic uniqueness of totality and scientists who study the Sun's atmosphere and magnetic fields during those times—know the path that the Moon's shadow will take across the surface many years in advance, and special tours to bring groups of astronomers and tourists to exactly the right places at exactly the right times have been organized since the 1700s.

Spectacular details were seen in the solar corona (the Sun's extended atmosphere) during the August 1, 2008, total solar eclipse over central Asia.

Examples of enormous (measuring many Earth diameters) coronal loops on the visible surface of the Sun. Loop diving, anyone?

(blocking) the visible disk of the Sun. Such "starshade" blockers—called coronagraphs by astronomers—don't work well on Earth (there is still too much scattered sunlight from the sky), but they have been used effectively on robotic solar-weather satellites since the 1970s. They are relatively easy to deploy from spaceliners, too, and tourists traveling all over the solar system have enjoyed viewing the beauty and subtle details of the Sun's atmosphere this way since the early days of space tourism. On your voyage close to the Sun, special large-scale occulting disks and special color filters will allow you to see extremely fine details on and above the Sun's visible surface as well as high up in its extended atmosphere. Take advantage of the opportunities to see the Sun like you've never seen it before.

CORONAL LOOP DIVE

Enormous magnetic fields emerge from the surface of the Sun and spread high-energy particles—the solar wind—out into the solar system, just like invisible magnetic field lines spread out from a bar magnet.

As the Sun rotates on its axis, however, those field lines become twisted and tangled. Sometimes they rip apart under the stress, and the broken end of the prominences—the high density parts of the field closest to the Sun—will loop back and reconnect itself to the surface. These so-called coronal loops are enormous structures, usually many times larger than the Earth, and they've been used to study the details of the Sun's magnetic fields for centuries. When they occur, they also provide the opportunity for a brand-new kind of thrilling but dangerous side excursion on your trip to the Sun. If the captain and crew believe that the timing is right and that environmental conditions are safe, and with the concurrence of the passengers, they can steer the ship directly toward and then *through* one of these loops. It's a dangerous, high-speed, headfirst dive almost to the surface of the Sun itself, and it's only been attempted (successfully) a few times to date. It will stress your spaceliner and its systems almost to the limits, but the thrill of the dive, the sheer awesome beauty of being surrounded by glowing high-energy solar energy, and then the high-gravity climb back away from the Sun is sure to be a once-in-a-lifetime experience.

HISTORY OF EXPLORING THE SUN

- **~2100 BCE:** Chinese astronomers observe and record sunspots

- **1543:** Copernicus revives the concept of a Sun-centered solar system, this time for good

- **1639:** First observed transit of Venus across the disk of the Sun, based on astronomer Johannes Kepler's predictions

- **1859:** First recorded observations of a solar flare

- **1868:** Discovery of the element helium from spectra of the solar corona during a total eclipse

- **1910:** Astronomers discover that the Sun is a typical star, like billions of others in our galaxy

- **1939:** Physicists discover fusion of hydrogen into helium is the source of the Sun's energy

- **1994-95:** *Ulysses* robotic space probe acquires data over the south and north poles of the Sun

- **1995:** *SOHO* satellite begins continuous stream of space-based solar imaging and monitoring

- **2004:** First samples of the solar wind returned to Earth by the *Genesis* robotic mission

- **2012:** *Voyager 1* becomes first space probe to leave the Sun's magnetic field

- **2060:** Several international robotic probes return first samples of Sun's lower atmosphere

- **2150:** Astronaut crew of *Helios 1* mission perishes in first attempt to sample a solar flare

- **2160:** Crew of *Helios 5* successfully acquires and returns solar photosphere samples to Earth

- **2210:** Thermal and radiation technology demonstrated for routine close-to-Sun flights

- **2218:** Solar System Tours Inc. begins offering tourist flights to the Sun

GETTING THERE

Currently only one company, Solar System Tours Inc., offers trips near the visible surface of the Sun, and so opportunities are limited and the cost is high (though rumors are that competition will come online soon). Accommodations and dining on the company's *Icarus* spaceliner are 5-star, so despite the inherent dangers in the trip (especially if a coronal loop dive becomes part of the excursion), passengers are afforded the highest possible levels of comfort in modern solar system travel. Total duration of each trip typically runs three or four weeks, depending on the specifics of the itinerary and the level of activity of the Sun. On the way there and back you can take part in lectures by solar astronomers and view the amazing detail of the Sun's surface and atmosphere through specially designed telescopes and eclipse-making coronagraphs operated by the crew.

THINGS TO DO, PLACES TO STAY

So far, options are limited to just one trip provider, perhaps because the space-tourism industry as a whole is waiting to see if there is enough customer demand to warrant additional spaceliners devoted to solar tourism, and if indeed Solar System Tours can maintain a spotless record of safety given the highly dangerous nature of the environment in which the company is operating. (The industry hasn't forgotten the lessons of the *Helios 1* tragedy from back in 2150.)

Still, the luxurious accommodations and spacious nature of the specially designed *Icarus* spaceliner provide ample opportunities for great dining, exercise, shows, lectures, and even a small spa that offers unique solar-powered hot-stone massage and hot yoga sessions.

LOCAL FLAIR

As you can imagine, there is still a bit of a "cowboy" culture associated with the risky nature of traveling so close to the Sun. Many of the pilots and crew of the *Icarus* previously worked on military space-vehicle test programs, or they were part of the speed-racing culture on Mercury. Get to know them, enjoy the dramatic stories (and tall tales) of their adventures at high speed or high temperature, but most of all, listen and observe them carefully throughout the trip to make sure you stay as safe as possible.

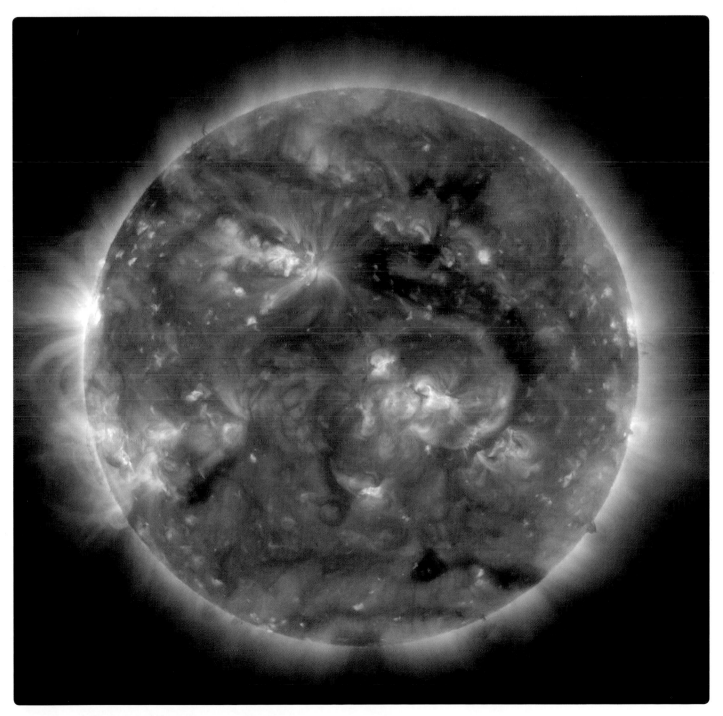

Using special telescopes and filters on your spaceliner, you can view the steady stream of high-energy particles that come out of the Sun.

9

TOURING THE MAIN ASTEROID BELT

Looking to venture beyond the Near-Earth Asteroids? Set your sights on the solar system's more than 30 million known asteroids (as of 2218) that orbit the Sun between Mars and Jupiter. Known as the Main Asteroid Belt, or just the Main Belt for short, this large, doughnut-shaped disk includes millions of small worlds made primarily of rock, metal, ice, or mixtures of those components. Generally, though, the rockier and more metallic bodies are closer to the Mars side of the Main Belt, while the icier worlds reside closer to Jupiter, reflecting the decrease in temperatures as they transition from the warmer climes of the inner solar system to the colder conditions of the outer solar system. Indeed, the Main Belt is often called the "gateway to the outer solar system," and the major bases that are found on the larger Main Belt Asteroids are frequently used as waypoints for travelers venturing to Jupiter and beyond, as well as major supply depots for those more distant destinations.

Opposite: A Ceres Travel Bureau poster welcoming visitors to the solar system's largest asteroid, the "Queen of the Asteroid Belt" and "Gateway to the Outer Solar System."

A size comparison of the three permanently inhabited Main Belt Asteroids with Eros, one of the largest Near-Earth Asteroids. For scale, Vesta is about 325 miles (525 kilometers) in diameter.

VESTA

PSYCHE

EROS

CERES

While there are millions of individual asteroids in the Main Belt, their total combined mass is only about 4 percent the mass of the Moon. A third of the mass of the Main Belt is contained in just one object, icy Ceres (the first asteroid discovered, in 1801, and the largest of all the asteroids). More than half of the Main Belt's mass is contained in Ceres plus the next three largest asteroids: Vesta, Pallas, and Hygiea. The largest metallic asteroid is Psyche, which is about 155 miles (250 kilometers) in diameter. Only about 60 asteroids are larger than 100 miles (160 kilometers) in diameter. Thus, like the Near-Earth Asteroids, most Main Belt Asteroids are very small worlds with very weak gravity. None of them have a permanent atmosphere, although a small percentage of them sometimes act like comets and "outgas" small amounts of water vapor or other gases to give them thin, temporary atmospheres.

There are as yet only a small number of permanently inhabited bases in the Main Belt, and the only ones that can currently accommodate tourists are found on Ceres, Vesta, and Psyche (the other bases are predominantly for workers in the mining and cargo-transport industries). Thus, your long-term stay and excursion options are limited in the Main Belt, although several spaceliners do offer flyby tours of a number of other small worlds. If you're planning to visit an asteroid in the Main Belt, you should prepare for:

Low Gravity: Like Near-Earth Asteroids and the moons of Mars, most Main Belt Asteroids are very small worlds with very little gravity. So prepare for low-gravity excursions on the surface, including on the bases. This means taking the time for the proper low-g training with your space suit as well as walking and docking methods for moving around on asteroids.

MAIN ASTEROID BELT FAST FACTS

Types of Bodies
Small rocky, metallic, or icy asteroids

✳

Distance from Sun
Most Main Belt Asteroids reside between the orbits of Mars and Jupiter, ~2.2–3.5 AU, or 195 million–325 million mi. (315 million–525 million km), from the Sun

✳

Distance from Earth
Closest approaches to Earth range from ~100 to 230 million mi. (~165 to 375 million km)

Travel Time from Earth
Varies from ~3 weeks to 1 year, depending on distance (if using slower spaceliner options, consider departing from Mars or one of its moons)

✳

Diameters
Largest (Ceres) is 600 mi. (960 km) in diameter; only ~60 are larger than 100 mi. (160 km) in diameter; tens of millions more are closer to ~328 ft. (100 m) in diameter

✳

Highlight
Some are rocky, some are metallic, some are icy; they are the gateway to many of the resources needed for solar system construction, life support, and exploration.

AVERAGE TEMPERATURES
(USING VESTA AS A TYPICAL EXAMPLE)

MAX. DAYTIME HIGH		MIN. NIGHTTIME LOW/SHADOWS	
°F	°C	°F	°C
27	-3	-306	-188

DON'T MISS . . .

VESTA

Vesta is a great first stop on a tour of the Main Belt. At an average diameter of 325 miles (525 kilometers), Vesta is the largest asteroid in the inner belt (see box on p. 85), and is the second-largest asteroid in the Belt overall. Robotic missions back in the early twenty-first century discovered that Vesta is more than just a typical primitive asteroid, however. Vesta is a small "protoplanet" that started down the path of becoming a full-fledged planet but never grew large enough. It did segregate, though, into a core, mantle, and crust, like planets do, and even had active volcanic eruptions on its surface early in

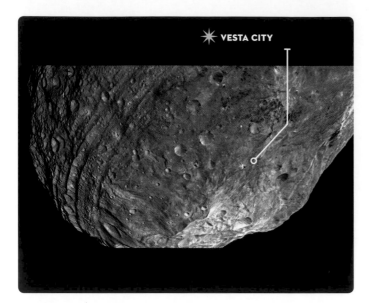

✦ VESTA CITY

solar system history (more than 4 billion years ago). As a result, Vesta's surface is covered in iron-bearing minerals similar to those that come out of volcanoes on Earth and other planets, making it a great source of iron and other extractable rocky building materials. Some of the easiest mining is at the large southern hemisphere impact crater called Rheasilvia. A bunch of Vesta's crust appears to have been ripped off from that impact, exposing a mixture of economically useful mantle and crustal materials.

Mining on Vesta began in the early 2100s as a way to supply iron and rocky materials to Ceres and other icy outer–solar-system outposts. By the mid-2100s, permanent residents working in and supporting the mines decided to incorporate Vesta City as an independently governed outpost. Happily, they decided to start welcoming tourists soon thereafter, recognizing curious and adventurous solar system travelers as an important source of additional income and jobs for native Vestans. Accommodations are not luxurious, but you will receive a friendly welcome and

Above: A close-up view of the southern hemisphere of Vesta. Vesta City is located along part of the rim of the enormous Rheasilvia impact basin.
Below: A plot of all the currently known asteroids, looking down on the solar system from above. The orbits of Mercury (inner circle) to Jupiter (outer circle) are noted, and the Main Belt Asteroids are the green dots between the orbits of Mars and Jupiter.

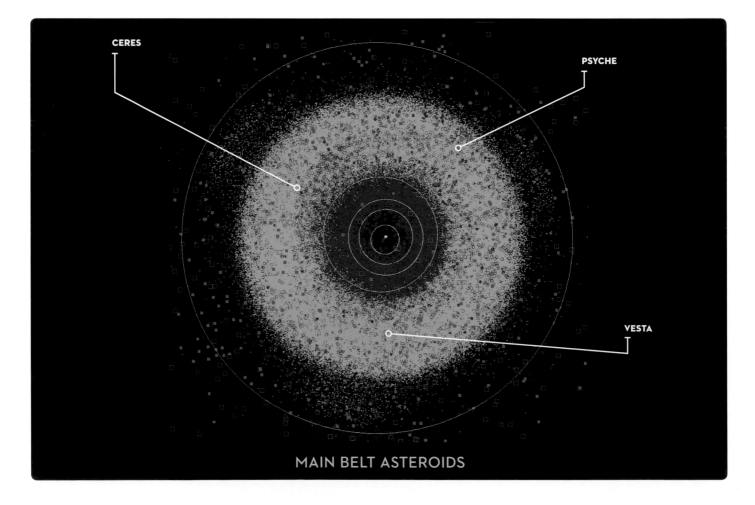

CERES

PSYCHE

VESTA

MAIN BELT ASTEROIDS

have opportunities to tour the mines and several other interesting geologic locations around the asteroid, and to dine cafeteria-style with hundreds of permanent residents of this easygoing colony.

CERES

Robotic space probes in the early twenty-first century discovered that Ceres, the largest asteroid in the Main Belt at 600 miles (965 kilometers) in diameter, is an icy middle-belt protoplanet that is peppered with salty mineral deposits, like the bright spots detected in Occator crater. Given the relative dearth of icy asteroids in the Main Belt, when people started contemplating moving out beyond Mars, Ceres quickly became the focus of entrepreneurial ice-prospecting and mining companies that saw a potential fortune to be made by turning Ceres into a major supplier and exporter of ice, salts, and other materials to destinations in the inner solar system, Main Belt, and outer solar system. Mid-twenty-first-century human missions to Ceres confirmed the quality and abundance of resources available

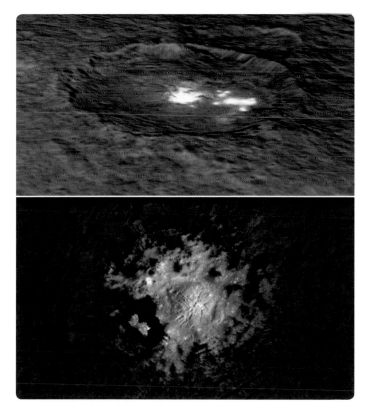

The 60-mile-wide (100-kilometer) Occator crater, viewed here in regional (top) and zoomed-in (bottom) images, is the home of Ceres Station, the first permanently established base in the Main Belt.

Examples of the spectacular metal spires of Psyche loom large approaching Psyche Station (along the rim of the small crater just below center).

there, and by the end of the century, a few thousand prospectors and other settlers banded together to found Ceres Station, the first incorporated base and spaceport in the Main Belt. They chose the rim of Occator crater as the site for the base, because of its easy access to huge deposits of bright ice and various mineral salts. Continuous mining and distribution operations began in the early twenty-second century.

Ceres Station remained a rather rough-and-tumble mining town for many decades, but the midcentury advent of Vesta City as a major Main Belt tourist destination convinced Ceres's leaders that they, too, could benefit from welcoming adventurous visitors. And they've been doing so for nearly 70 years now, offering a bit more "Wild West" flare than on Vesta, but similar sorts of relatively modest accommodations, dining options, and excursions. Particularly lovely are the tours of the Occator Salt Mines, which take visitors down through a labyrinth of natural and human-made subterranean caverns where salt has been extracted for more than a century. Parts of the caverns have been preserved in their natural form, sealed, and filled with breathable atmosphere, allowing visitors to bounce around gingerly (under only 3 percent of Earth's gravity) among the spectacular crystalline cave structures. Plans are underway to add these spectacular parts of the Occator Salt Mines to the Interplanetary Park System, hopefully preserving them for visitors indefinitely.

PSYCHE

Late-twentieth-century radar observations from Earth identified the outer belt asteroid Psyche as the largest metallic asteroid in the solar system. Follow-up twenty-first-century robotic orbiter and sample-return missions confirmed the composition and revealed a variety of unique and bizarre landforms on this 155-mile-diameter (250-kilometer) world. Psyche, like Vesta and Ceres, appears to be a protoplanet (a small world that segregated into core, mantle, and crust while on the path of growing into a larger planet), but it underwent one or more catastrophic collisions with other asteroids early in solar system history that stripped off its crust and mantle, leaving just Psyche's metallic inner core exposed to space. Thus, Psyche is a metal world, dominated by iron and nickel but also loaded with huge deposits of other rare precious metals. The prospectors and mining speculators of the mid-twenty-first century went wild about the possibilities.

By the mid-2100s, claims were being widely staked by entrepreneurs and early settlers on Psyche, and within a few decades there were enough people living and working (and making lots of money) there to establish a permanent base, Psyche Station, as a spaceport and coordination center for the growing metals-extraction and export industry. While the bottom soon fell out of the Psyche mining market, locals decided recently to place new bets on a high-end tourist economy for Psyche. It appears to be paying off, with luxury accommodations, dining, and novel entertainment and excursion options attracting visitors from all over the solar system to this new outer-belt hot spot (see Local Flair below).

GETTING THERE

The Main Belt is at the intersection of the inner and outer solar systems. Because of the long distances to these outposts, travel times are significantly longer and the frequency of spaceliner flights is significantly less than to inner-solar-system destinations.

The fastest possible trips can get you from Earth to the inner Main Belt (to Vesta City, for example) in as little as three weeks, if the trip is timed for the one time a year when Earth and Vesta are as close as they get in their orbits. The distances are astronomical, and because of the competition for the rare launch opportunities, so are the prices, which typically run from five to ten times the cost

of a typical high-speed trip to Mars. Routine high-speed spaceliner service doesn't yet run to middle or outer Main Belt destinations, though special tourist runs to Psyche are now being scheduled, and there are rumors of several companies starting trips to Ceres, and possibly even high-speed runs from Mars to all major Main Belt destinations, within a decade.

Traditional (slower) propulsion spaceliner flights are more common to Main Belt destinations, though they are also less frequent than their inner-solar-system counterparts. Travel times from Earth can easily mount to six, nine, or twelve months depending on how far into the Main Belt you plan to travel. If you are planning to visit the Main Belt as part of an extended vacation (or a one-way trip or retirement), consider departing from Mars, which is more than halfway to the Main Belt. Several companies also offer special tours of a variety of the smaller Main Belt Asteroids, where local mining operations welcome orbital tourists.

THINGS TO DO, PLACES TO STAY

As you might expect given the relatively small number of permanent settlements currently established in the Main Belt, accommodation, activity, and entertainment choices are very limited compared to most inner-solar-system destinations. Indeed, only recently have luxury 5-star accommodations become available in the Main Belt, on Psyche, where colonists are working hard to attract tourists deep into the Belt after the precious-metals market collapsed. The twin Id and Ego resorts and spas on Psyche (one adult-themed, the other family-themed) just opened to great fanfare and wonderful restaurant reviews. Otherwise, tourist accommodations are limited to decent (3-star) hotel and dining options for visits to both Ceres and Vesta, or to the usual range of spaceliner rooms for tours of other asteroids.

Tethered hiking tours are available on all three permanently settled large Main Belt Asteroids, offering great vistas of the geologic wonders of these small worlds. Indoor and underground tours of mining facilities are also available, with the ice production plants on Ceres

HISTORY OF EXPLORING MAIN BELT ASTEROIDS

- **1801:** Discovery of Ceres (1st asteroid discovered), from Palermo Observatory, Italy

- **1807:** Discovery of Vesta (4th asteroid discovered), from Bremen, Germany

- **1852:** Discovery of Psyche (16th asteroid discovered), from Naples, Italy

- **1868:** 100th asteroid discovered by astronomers

- **1923:** 1,000th asteroid discovered by astronomers

- **1951:** 10,000th asteroid discovered by astronomers

- **1982:** 100,000th asteroid discovered by astronomers

- **1991:** First asteroid flyby by a spacecraft: 951 Gaspra studied by the *Galileo* spacecraft

- **2011–12:** *Dawn* spacecraft becomes first to orbit a Main Belt Asteroid (Vesta)

- **2014–17:** *Dawn* spacecraft studies the largest asteroid, Ceres, from orbit

- **2026:** *Psyche* spacecraft studies the largest metallic asteroid, Psyche, from orbit

- **2028:** 1,000,000th asteroid discovered by astronomers

- **2045:** First robotic Main Belt Asteroid sample return (*Psyche-2* mission)

- **2056:** First human mission to a Main Belt Asteroid (*Ceres-1* orbiter and lander)

- **2080:** 10,000,000th asteroid discovered by astronomers

- **2099:** Ceres Station, first permanent Main Belt base, established in Occator crater

- **2112:** Continuous ice and salt extraction and mining begins on Ceres

- **2146:** Vesta City established as first Main Belt rock/iron mining colony, in Rheasilvia crater

- **2150:** First tourist excursions offered to Ceres and Vesta

- **2180:** Psyche Station established as first Main Belt precious-metals mining colony

- **2218:** Psyche Station becomes tourist and research outpost after crash of precious-metals market

being the most extensive and awe-inspiring (including, for example, a substantial network of caves with stunningly large crystalline stalactites and stalagmites). Traditional rock and iron processing plants on Vesta provide interesting guided tours of modern low-*g* mining technologies, and those interested in mining history can even visit the now-defunct precious-metals mining stations on Psyche.

LOCAL FLAIR

The discovery in the 2020s of substantial deposits of iron, nickel, platinum, palladium, uranium, and other rare precious metals exposed on the surface of Psyche quickly made it a target for speculators and miners, but it took until the late 2100s to raise enough capital, and convince enough colonists, to move there and begin mining operations. For more than 30 years, the "Psychotics" (as the locals call themselves) were easily able to extract enormous amounts of extremely rare metals and ship

Enjoy a spectacular sunrise on Vesta as a shuttle takes you to your tour of one of the asteroid's many iron-mining camps.

them back to Earth, the Moon, and Mars for wide use at cut-rate prices. The 500 or so permanent residents who composed the Psyche Metals Collective became some of the richest and most eccentric people in the solar system, creating spectacular private residences on Psyche that are counted among the most remarkable low-g mansions and estates in the solar system.

However, the precious-metals economy has essentially collapsed in the past 10 years, as companies closer to home became able to capture and quickly extract precious metals from small, Near-Earth Asteroids in space, for significantly less than the cost of shipping from Psyche. Undeterred, members of the collective have set out to use some of their immense wealth to create a tourist economy on Psyche, converting several estates into resorts and spas catering to the high-end tourist market. Time will tell if the experiment is able to attract a sustainable number of tourists.

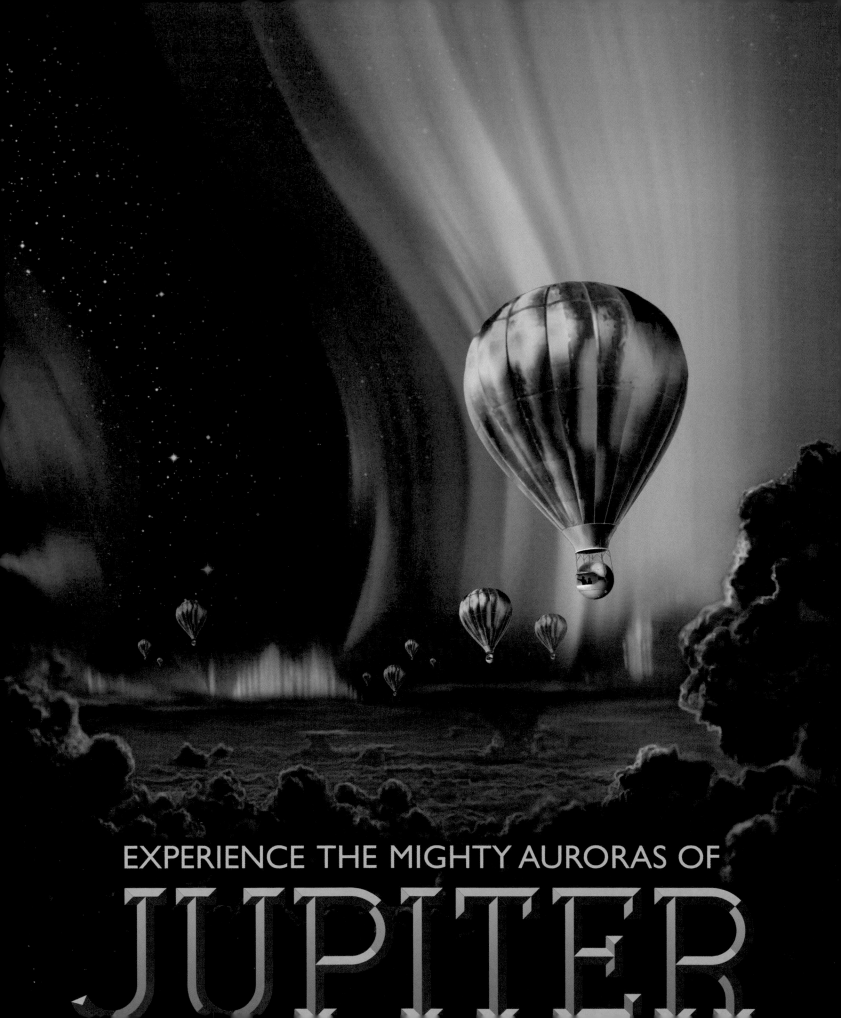

EXPERIENCE THE MIGHTY AURORAS OF
JUPITER

EXPLORE JUPITER AND THE GREAT RED SPOT

Jupiter is often called the King of the Planets, but even that grand moniker falls short when you look at the numbers: Jupiter is over 2.5 times more massive than all of the other planets, moons, asteroids, and comets in the solar system *combined*. Twenty-three Earths could fit across the diameter of mighty Jupiter, and more than a thousand Earths could fit inside this massive planet. Indeed, Jupiter is so massive that it almost became a second star in our solar system. Like the Sun, Jupiter was born with about 75 percent hydrogen and 25 percent helium, but it has only 0.1 percent the mass of the Sun, and so the temperatures and pressures inside Jupiter are not high enough to initiate nuclear fusion. Thus, Jupiter remains in the realm of the planets rather than the stars.

Opposite: Whether by airhotel, shuttle, or balloon, the mighty auroras of Jupiter are a sight to behold.

A full-disk view of Jupiter, the King of the Planets. The Great Red Spot (just below right of center) is nearly three times the diameter of the Earth.

An ultraviolet view of intricate aurora structures over the north pole of Jupiter.

Jupiter is the largest so-called gas giant planet (a planet composed mostly of the gases hydrogen and helium and without solid surfaces) in our solar system. Its visible "surface" is not a solid surface at all, but rather consists of roiling, turbulent layers of clouds and gases colored by exotic molecular compounds like methane, ethane, ammonium hydrosulfide, and phosphine. Intense winds from the heat of Jupiter's interior organize the cloud layers into a variety of bands called zones and belts, and turbulence along the boundaries of these regions creates enormous hurricane-like storm systems. The largest of these is the Great Red Spot, a huge, swirling, high-pressure system that ebbs and flows from one to three times the size of Earth, and that has been raging in Jupiter's southern hemisphere since at least 1665, when it was first noticed by astronomers who were using some of the earliest astronomical telescopes.

So while you can't walk on Jupiter, you can sign up to fly through those magnificent Van Gogh–like clouds and storms, cruise among the magnificent northern-lights auroras on the planet's night side, or float lazily in comfort and luxury among the warm, billowy water-vapor clouds just below the upper-haze layers, dreaming fancifully of cloudy days back on your own home planet.

BEFORE YOU GO

Jupiter is visually beautiful, but the environment is highly stressful for human visitors. Dangerous radiation levels, high gravity, crushing atmospheric pressures, and powerful winds are among the challenges that you, your fellow travelers, and your spaceliner crew will face when encountering the King of the Planets. When you start planning your trip, you'll want to prepare for the following:

☢ **Extreme Radiation:** Jupiter's rapidly spinning metallic core creates the strongest magnetic field in the solar system (besides the Sun's), and that field creates enormous amounts of high-energy radiation, which is extremely dangerous to life-forms like us, as well as to

JUPITER FAST FACTS

Type of Body
Planet (gaseous), about 318 times
more massive than Earth

✳

Distance from Sun
Average is ~5.2 AU, or 484 million mi.
(778 million km)

✳

Distance from Earth
Ranges from 391 million to 580 million
mi. (630 million to 930 million km)

Travel Time from Earth
Varies from about a month to a year,
depending on distance

✳

Diameter
Approximately 87,000 mi.
(140,000 km), or about 23 times
Earth's diameter

✳

Highlight
Fantastically colorful and energetic
storms, like the Great Red Spot!

AVERAGE PRESSURE AND TEMPERATURES

REGION OF THE PLANET	AVERAGE PRESSURE	AVERAGE TEMPERATURES	
	(EARTH'S SURFACE = 1.0)	°F	°C
Lower stratosphere: Tops of the highest clouds and hazes	0.1	-244	-153
Earth's surface pressure level (31 mi., or 50 km, below the haze)	1.0	-154	-103
Bottom of water-ice cloud layer (62 mi., or 100 km, below the haze)	10	80	27
Metallic hydrogen mantle (19,000 mi., or 30,000 km, deep)	3,000,000	14,000	7,700
Center of Jupiter's rocky/metallic core	12,000,000	45,000	25,000

electronic equipment. Luckily, the same kinds of advances in radiation-shielding technology that make it possible to travel very close to the Sun also enable travel very close to Jupiter. So you'll be protected, but during the most intense radiation periods you might have to take shelter for several hours in the special water-lined safety pods built into every Jupiter-bound spaceliner. As always, pay attention to the safety drills!

High Gravity: Getting close to Jupiter will weigh you down—literally—because the force of gravity at Jupiter's cloud tops is about 2.5 times Earth's gravity. As you skim by those glorious clouds, your spine will shrink and you'll feel dramatically overweight and weak. While special spinning sections of Jupiter-bound spaceliners can help to compensate for this effect, some people have trouble maintaining their equilibrium for long periods of time in a spinning environment. People who take time to acclimatize to the increased gravity (using special exercise equipment, for example) and/or to the spin report good experiences dealing with the high gravity, especially for short flyby visits to the planet.

As you get closer to Jupiter's cloud tops—as in this close-up of the Great Red Spot from NASA's *Juno* spacecraft in 2017—you'll be presented with a dizzying vista of swirling and colorful clouds and a storm system that seem to be right out of an impressionist painting.

Extreme Pressure: Now on offer are new kinds of voyages that descend below the cloud deck, into the depths of storm systems like the Great Red Spot. Be warned, however, that travelers will have to deal with the secondary effects of being immersed in a high-pressure environment. For example, descending to the airhotels among the water-vapor clouds about 62 miles (100 kilometers) deep in Jupiter's atmosphere places people in a region where the pressure is 10 times the surface pressure of Earth. This is comparable to diving about 230 feet (70 meters) under water. While the structure of your hotel and spaceliner is designed to withstand this pressure difference, pressure inside the vessel is often

raised somewhat to help compensate. This can cause dizziness for some, and those susceptible to pressure changes may also experience symptoms comparable to the decompression sickness that ails deep-sea divers. All spaceliner companies offer pre-voyage pressure testing and acclimatization training, so you can prepare for your deep dive into Jupiter.

DON'T MISS . . .

It typically takes many months to get to Jupiter, even on the fastest spaceliners available, and so to make the voyage worthwhile, you're going to want to spend some quality time around—and within—this giant planet. Luckily, you now have several airhotel choices, and the variety of spaceliner tours will afford you some of the most spectacular vistas in the solar system.

THE GREAT RED SPOT

If you were to pick one spot that has defined everyone's impression of Jupiter since the invention of the telescope, it would be this one. *Great* because it is the largest storm system in the solar system—at its largest, three Earths could fit inside! *Red* because of the garish colors that molecules such as ammonia, ammonium hydrosulfide, and acetylene create as they mix and rise and form clouds in the storm's updrafts. And *Spot* because it's been there at that same southerly latitude—spinning on its axis every five days—since *at least* the 1660s, and probably much longer than that. It has grown and shrunk over the centuries, become more and less red, but still just continues to churn.

Since the late 2100s, people have been able to get a front-row seat to this magnificent show by staying in the Great Red Spot airhotel, a luxury resort and spa that floats lazily above the reddish cloud tops. From there, you can see the spectacular three-dimensional nature of the Spot, parts of which rise like a gargantuan thunderstorm above the surrounding cloud layers. Huge windows (and rocking chairs) let you sit and stare at the magnificent turbulent patterns of eddies and vortices that swirl around you,

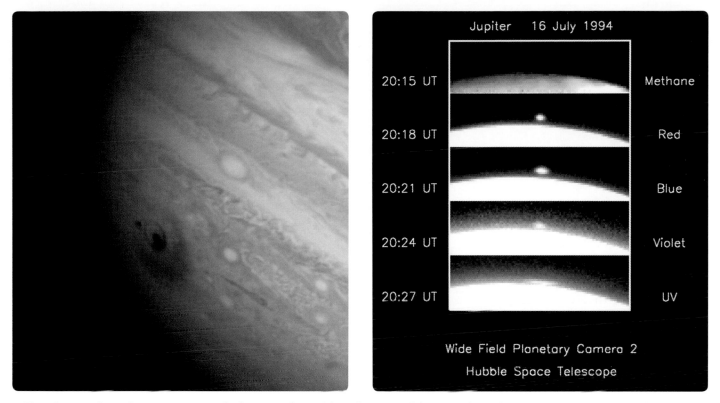

The airburst explosion (time sequence at right, from top to bottom) from the impact of the comet Shoemaker-Levy 9 into Jupiter's atmosphere in 1994 left enormous Earth-size scars and blemishes (left) on the giant planet for months.

changing colors as clouds rise and fall and as the sunlight varies over the course of the very short (about five hours) Jovian daytime. Accommodations are high-end, and the dining (featuring ingredients from all over the solar system) is widely regarded as the best in the Jupiter system. Plan to stay—and become mesmerized—for a week, or a month if you can afford it.

AURORA VIEWING ON JUPITER'S NIGHT SIDE

Visitors to Earth's polar regions know that one of the most spectacular sights to take in is the nighttime northern (or southern) lights. Earth's auroras are beautiful, but they pale in comparison to the brighter, faster, and more colorful polar lights created by Jupiter's 14-times-stronger-than-Earth's magnetic field.

Hotel Zeus, a specially located airhotel, orbits high above Jupiter's polar regions, providing 3- and 4-star accommodations for travelers who want to "take in the

lights" of the Jovian aurora borealis. The ship stays in sight of the night side of Jupiter at all times, enabling constant viewing (and, for the researchers on board, scientific study) of the northern and southern lights below. Shuttle excursions and even specially equipped balloon flights are offered for rides down into the auroral zone, where you can be magically surrounded by shimmering reds, greens, and yellows. You'll also see plenty of lightning and maybe even enormous, strangely shaped, colorful "sprites"—electrical discharges from cloud to cloud, high above convective thunderstorms. It will be a bumpy but an immensely memorable flight!

WITNESS AN IMPACT!

Starting with the famous 1994 impact of Comet Shoemaker-Levy 9 into the clouds of Jupiter, astronomers have learned that comets and asteroids are constantly hitting the giant planet. Most impact events are small and hardly noticed; larger events like the one back in

WHAT CAUSES JUPITER'S AURORAS OR "NORTHERN LIGHTS"?

In the same way that the field lines from bar magnets line up iron filings in that classic high-school science demonstration, the magnetic fields of the planets "line up" high-energy particles and steer them in toward the source of the magnetism. Thus, when solar wind particles—high-energy atoms and protons streaming off the Sun—feel the attraction of a magnetic planet like the Earth or Jupiter, those particles are accelerated toward the planet's surface (or cloud tops). They are focused toward the polar regions, because that's where the magnetic field lines converge. In the process, the electrons that make up those particles gain energy and become excited (ionized). Ions that escape the field then give back all that energy by releasing visible-light photons, creating the shimmering reds, greens, and yellows of the aurora. The rapid changes in the shapes and colors of the aurora reflect both the highly variable nature of particles streaming off the Sun, as well as the highly variable nature of the planet's magnetic field. The end result is a colorful, dynamic, and ultimately beautiful display of space physics in action.

1994 only occur a few times a century. But the relatively common smaller impacts (asteroids or comets from 30 to 165 feet, or 10 to 50 meters, in size) have spurred another burgeoning tourist sector on Jupiter—impact hunting. Visitors sign up for a special weeklong spaceliner cruise, board the vessel, and then . . . hang out. Astronomers on board track small objects approaching Jupiter, and when they notice one of them on a collision course, they steer the ship to witness the event. Small impacts typically occur a few times a week, so you won't have to wait long for the show.

And what a show it is! You can experience the same sense of shock and awe that astronomers did back in 1994, when they realized that even a small object—like a chunk of a comet nucleus only a few football fields across, but traveling at super-high speeds—can release more energy than all the nuclear weapons on Earth. Your crew will position you a safe distance away and give you special glasses to watch the *trinity moment* of impact, when all that energy is focused into shock waves and both the impactor and the atmosphere are subjected to high-temperature vaporization. Exotically colored molecules are dredged up from much deeper atmospheric layers, and the resulting scars and blemishes in the cloud tops will persist for months. As the temperatures cool, your crew will take you on a closer tour of the blast zone. It's a trip that is sure to have a lasting impact!

SLEEP BELOW THE CLOUDS

Jupiter's cloud tops are icy cold because the planet is so far from the Sun. But as you descend deeper into the atmosphere, the increasing pressure results in increasing temperatures. Because of this effect, somewhere between about 30 and 60 miles (50 and 100 kilometers) below the cloud tops you'll reach a layer of the atmosphere where temperatures and pressures are relatively Earthlike. Giant billowing cumulus clouds of water vapor form against deep-blue skies in some of these zones, and even though there is no land or ocean to be seen below, visitors nonetheless report a sense of feeling like they are back on Earth, floating among the familiar clouds of our home planet.

A stormy scene on Earth? No, it's the view from 30 miles (50 kilometers) below the upper-haze layers of an Earthlike zone of modest temperatures and pressures in Jupiter's atmosphere.

That sense of experiencing something familiar in such an otherwise alien landscape has recently motivated some entrepreneurs to establish several "Nimbus" airhotels in these Earthlike zones of Jupiter's atmosphere. While you can't breathe the air, you can don a relatively simple air mask and stroll out on the deck and veranda in your shirtsleeves. For many, it will be the first time they've experienced the feeling of truly being outdoors—without the confines of a space suit—for a year or longer. As yet, these establishments are only offering basic accommodations, and dining and entertainment choices are limited. Still, the draw of somehow, strangely, feeling "at home" is proving to be a successful tourist attraction, and so the accommodation and other choices will hopefully increase in the future.

A STAR OR NOT A STAR?

Is Jupiter a giant planet or a small, failed star? Astronomers have discovered that in order for a star to be a star, its central core temperature and pressure have to be high enough to make four hydrogen atoms fuse into one helium atom—releasing a tiny amount of energy as light and heat in a process that balances gravity's attempt to compress the star further. The smallest size an object can be and still have such nuclear fusion is about 8 percent the mass of the Sun. That is still about 80 times the mass of Jupiter, however, so while it is crazy hot and dense in Jupiter's core, it's not hot and dense enough to fuse hydrogen. Close, but no sunshine, Jupiter.

HISTORY OF EXPLORING JUPITER

- **1610:** Galileo is the first to observe Jupiter and its orbiting moons through the telescope
- **1665:** Astronomers Robert Hooke and Giovanni Cassini discover Jupiter's Great Red Spot
- **1955:** Radio astronomers discover Jupiter's incredibly strong magnetic field
- **1973:** *Pioneer 10* performs first spacecraft flyby of Jupiter
- **1978:** Jupiter's auroras discovered by the *International Ultraviolet Explorer* (*IUE*) Earth-orbiting space telescope
- **1979:** *Voyager 1* and *Voyager 2* acquire first high-res images of Jupiter, its moons, and rings
- **1994:** Comet Shoemaker-Levy 9 slams into Jupiter
- **1995:** *Galileo* becomes first spacecraft to orbit Jupiter and first probe to enter Jupiter

- **2016:** *Juno* orbiter mission maps the gravity and interior structure of Jupiter
- **2041:** First robotic sample-return to Earth of Jupiter's atmospheric gases
- **2078:** First human mission to Jupiter (Mars launch, Jupiter flyby, then return to Mars)
- **2120:** First "deep probe" robotic mission to sample Jupiter's metallic hydrogen interior
- **2153:** Orbital Magsat network begins to beam electricity to bases on Jupiter's moons
- **2180:** First cloud-top airhotel established on Jupiter (Great Red Spot Resort)
- **2200:** First deep airhotels established in Jupiter's Earthlike pressure and temperature zones
- **2218:** New round-trip excursions being offered into the deep atmosphere

GETTING THERE

Jupiter is just at the inner edge of the outer solar system, but the time it takes to get there will still remind you that the solar system is enormous, and that the distances to the outer planets are—truly—astronomical. The fastest modern propulsion spaceliners can get you there from Earth in about a month (or about two weeks from Mars), but you'll have to time your departure for when the planets are aligned just right, and you'll pay premium prices. Older, more traditional (and more affordable) spaceliner services can get you to Jupiter in about a year (from Earth) to about six months (from Mars). Because of the long travel times, there are far fewer flight opportunities to and from Jupiter compared to inner-solar-system destinations, so be sure to book ahead.

THINGS TO DO, PLACES TO STAY

The long transit times to and from Jupiter will make your spaceliner your home for most of your trip, so be sure to choose accommodations with shipboard entertainment and dining options that will work for you for a long voyage. Once at Jupiter, you can choose to either keep your accommodations on the spaceliner (which will usually go into orbit and offer tours and excursions from there) or transfer directly to an airhotel. Look for package deals, and remember that there is shuttle service between most of the major venues. Entertainment and dining options vary widely among the airhotels, from 5-star Great Red Spot options to midlevel Zeus auroral viewing options to the closest thing to camping on Jupiter, staying at one of the basic Nimbus airhotels below the upper-cloud decks.

All of the spaceliners and airhotels will have entertainment, recreation, and excursion options, ranging from spa services to bars and nightclubs to research-lab visits and lectures. Photography is a common pursuit for many Jupiter-bound travelers, and you'll find ample opportunities to learn and practice the right techniques for capturing fleeting auroras or subtle details in the planet's thin, dark rings.

LOCAL FLAIR

Most passengers on long spaceliner voyages to the outer solar system end up becoming good friends with their spaceliner pilots and crew. You'll probably learn that many crew members on the traditional propulsion "long haul" runs are traveling with their families, and that their spouses and even older children are perhaps part of the crew or the dining and entertainment staffs. To them, the ship *is* their home, and some of the children you meet may never have felt the pull of old-fashioned gravity on an actual planet or moon. Take their advice on how best to experience all that Jupiter has to offer.

Hop on a spaceliner and witness the impact firsthand as a small asteroid or comet slams into the coulds of Jupiter!

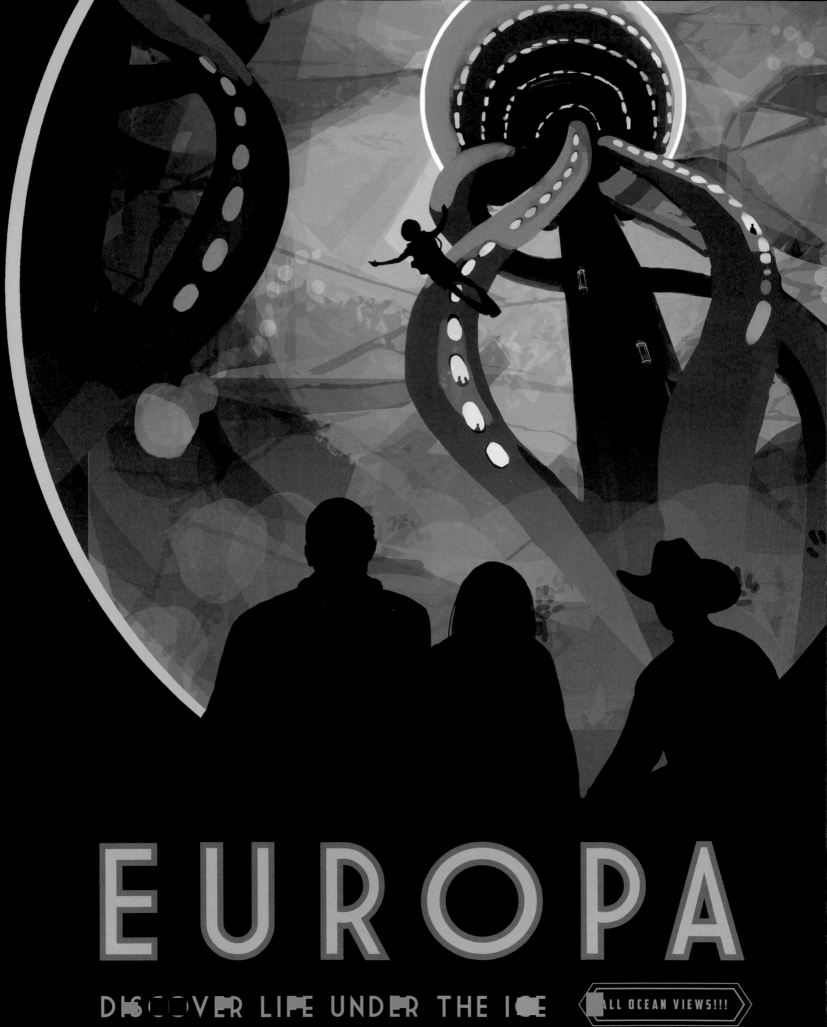

EUROPA

DISCOVER LIFE UNDER THE ICE   ALL OCEAN VIEWS!!!

VISITING EUROPA AND THE MOONS OF JUPITER

Back in 1610, Galileo's discovery of four star-like moons orbiting the giant planet Jupiter was revolutionary because it helped end the idea that Earth was the center of the universe. The discovery and subsequent exploration of what have since become known as the Galilean satellites of Jupiter—the large moons Io, Europa, Ganymede, and Callisto—have been revolutionary in other ways as well, especially because these worlds have turned out to be geologically, hydrologically, and perhaps even biologically active. At least two of these moons, Europa and Ganymede, are rocky and icy bodies that have liquid-water oceans under their icy outer shells. One, Callisto, is mostly icy but has a slushy liquid-water layer in its subsurface as well. And one, Io, is an entirely rocky body with the highest number of active volcanic eruptions of any place in the solar system. What Galileo had discovered turned out to be a fascinatingly diverse and interesting mini solar system of its own, orbiting around Jupiter.

Opposite: Submersible research stations in Europa's ocean are now starting to take on adventure-science tourists. Come aboard, and become a part of solar system history.

The most exciting of these worlds has turned out to be Europa. Of Jupiter's four large moons, it is the second farthest from the planet and is about 10 percent smaller than Earth's moon. There's no atmosphere on Europa, and the surface consists of a super-smooth shell of water ice, which is fractured into thousands of icy plates that appear to move slowly, relative to one another. The motion of these plates was one of the first hints of a deep, liquid-water ocean below that "sea ice." Indeed, based on data from robotic flyby, orbiter, and lander missions, planetary scientists in the late twentieth and early twenty-first centuries discovered that Europa's ocean has nearly three times the volume of all of Earth's oceans combined. It's

saltwater, suggesting prolonged direct contact between the water and the moon's rocky crustal materials. Organic molecules (complex chains of carbon, hydrogen, nitrogen, and other atoms) have been discovered in cracks on the surface and in samples of the ocean itself. And the persistence of the ocean over geologic time means that there must be an internal heat source keeping the water from freezing solid at Europa's great distance from the Sun. Liquid water, organic molecules, and heat sources are the ingredients required for a world to be habitable to life as we know it, and Europa is exciting because it is one of the few places in the solar system besides Earth that meets all of those requirements.

EUROPA STATION

NORTH
POLE

SOUTH
POLE

EUROPA STATION AND EUROPA'S POLES

EUROPA FAST FACTS

Type of Body
Planetary moon; icy exterior, deep subsurface ocean, rocky mantle and core

✳

Distance from Sun
Average is ~5.2 AU, or 490 million mi. (790 million km)

✳

Distance from Earth
Ranges from 391 million to 580 million mi. (630 million to 930 million km)

Travel Time from Earth
Varies from about a month to a year, depending on distance

✳

Diameter
3,100 km (1,900 miles), or about 10% smaller than Earth's moon

✳

Highlight
Europa's deep, salty ocean has almost three times as much water as all of Earth's oceans combined!

AVERAGE TEMPERATURES

SATELLITE	DAYTIME		NIGHTTIME		EXTREMES / SPECIAL LOCATIONS		
	°F	°C	°F	°C		°F	°C
Io	-225	-143	-297	-183	Lava flows	3000	1650
Europa	-256	-160	-364	-220	Ocean water	32-39	0-4
Ganymede	-234	-148	-315	-193	Ocean water	32-39	0-4
Callisto	-218	-139	-315	-193	Equator, noon	-162	-108

BEFORE YOU GO

Europa and the other large moons of Jupiter are exciting travel destinations, but when you start planning your trip, you'll want to prepare for the following challenges:

Extreme Radiation: The moons of Jupiter are bathed in high levels of dangerous radiation from the planet's enormous magnetic field. Spaceliners, shuttles, and surface and subsurface habitats are properly shielded, but you will still have to be trained to use radiation safety pods in case of intense radiation outbursts.

Extreme Temperatures: Europa and the other moons of Jupiter are a long way from the Sun, and they don't have insulating, warming atmospheres. Thus, even the warmest daytime temperatures are still frigid (around -240 degrees Fahrenheit, or -150 degrees Celsius), and nighttime temperatures plummet to a low of around -330 degrees Fahrenheit (-200 degrees Celsius), only about 130 degrees Fahrenheit (70 degrees Celsius) above absolute zero! Even the "warmer" oceans under the ice shells of Europa and Ganymede are only barely above the freezing point of water. These low temperatures mean that excellent thermal insulation is required in ships and habitats, as well as space and diving suits, and special training is needed to recognize and treat the effects of space hypothermia. Isolated lava lakes and lava flows on volcanically active Io are as hot as molten rock, in contrast, presenting different but just as challenging kinds of dangers.

Above: The Galilean satellites of Jupiter. Left to right, from closest to farthest from Jupiter, are Io, Europa, Ganymede, and Callisto. For scale, Io is about the same size as Earth's moon, and Callisto is about the same size as Mercury.

Below Right: This 100-mile-wide (160-kilometer) region of Europa shows the surface ice shell divided into thousands of small plates. Seawater from below gets to the surface between the plates, forming ridges colored red from salts left behind after the water evaporates.

DON'T MISS . . .

Numerous tourist opportunities focused on education, entertainment, and even active "citizen science" research have come online on Europa and the other large moons of Jupiter over the past few decades.

LIFE UNDER THE ICE?

While there are several dining, entertainment, and accommodation options available under the radiation-protecting and thermally insulated dome structure of Europa Station, the most exciting tourist activities are now occurring far beneath the dome, within Europa's deep and magnificent ocean. Scientists in submarine research stations have been studying and searching for life in that ocean since the mid-2100s, and now they are accepting small groups of tourists as a way to help defray their research expenses. Safe from the frigid temperatures and sizzling radiation of the surface and built for deep-sea exploration, these stations can float

to different levels of the ocean, or descend all the way to the ocean floor, making for unforgettable adventures to uncharted lands. As a part of the crew, you (and your kids) can participate directly in research and discovery on Europa, helping to choose destinations, select sampling areas, and oversee some of the onboard lab experiments that require human supervision or feedback. Life hasn't yet been discovered there, but scientists have only just begun the systematic search. While accommodations and food choices are modest, it's still a spectacular opportunity to be part of something truly historic—searching for extraterrestrial life in a spectacularly habitable environment.

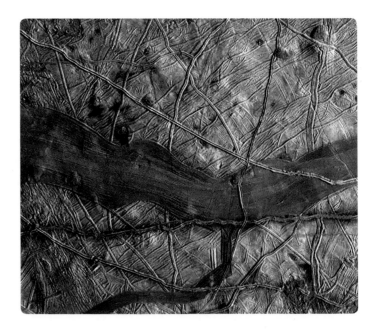

IO

Jupiter's innermost moon, Io (pronounced EYE-oh), which is about the same size as Earth's moon, is one of the most violent places in the solar system. Volcanic vents and cones across the surface spew out sulfur- and silicate-bearing ash and lava in a near-constant state of eruption. Sulfur changes color dramatically with temperature, leading to the gorgeous palette of white, yellow, orange, red, and black hues that cover and constantly renew the moon's surface. This volcanic wonderland has become a major outer-solar-system geologic research outpost, and Io Volcanic Observatory, located under a dome not far (but a safe distance) from the large volcano Loki Patera, is the center of activity. Io Station offers a few dozen 3-star hotel-style rooms for visiting tourists. You can dine with the science staff, attend public lectures and academic symposiums, and accompany your geologist guides on several hikes to the nearby lava lake and fire-fountaining vents. Hot times await!

GANYMEDE

Ganymede is the largest moon in the solar system, and it is more than 20 percent larger than the planet Mercury. Like Mercury, it has a core, mantle, and crust, but it also has complex plate-tectonics-like surface geology. It is also the only moon with its own magnetic field. Unlike Mercury, however, it is mostly made of ice and rock instead of rock and metal. Ganymede also sports a subsurface liquid-water ocean, but this lies beneath a much thicker icy shell than Europa's ocean, and so has yet to be reached by either robotic or human visitors. Still, Ganymede Station is an active research base where scientists study that moon's ocean and interior remotely (mostly using seismic waves, like geophysicists on Earth),

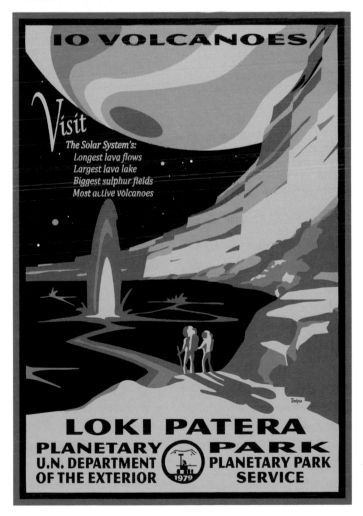

Top Left: Jupiter's colorful clouds provide a wonderful backdrop to spectacularly colored volcanic eruptions on Io.

Bottom Left: Volcanologists at Io Station can take you on some amazing hikes to view firsthand the most active eruptions in the solar system.

HISTORY OF EXPLORING EUROPA AND THE MOONS OF JUPITER

- **1610:** Galileo discovers Io, Europa, Ganymede, and Callisto through his telescope

- **1979:** *Voyager 1* and *Voyager 2* acquire first resolved images of the Galilean satellites

- **1995:** *Galileo* Jupiter orbiter acquires high-resolution maps of these moons

- **2013:** Plumes of water ice detected on Europa from *Hubble Space Telescope* data

- **2026:** NASA's *Europa Clipper* mission discovers conclusive proof of Europa's ocean

- **2029:** First robotic lander on Europa

- **2032:** European Space Agency's (ESA) *Jupiter Icy Moon Explorer* mission proves existence of Ganymede's ocean

- **2045:** First robotic Io orbiter and lander

- **2058:** First robotic Ganymede lander and rover mission

- **2066:** First robotic Callisto lander

- **2085:** First successful robotic submarine mission into Europa's ocean

- **2098:** First human landings on Jupiter's moons (first Ganymede, then Europa)

- **2115:** Europa Station's domed, surface research habitat established

- **2126:** Both Callisto Station and Ganymede Station research habitats established

- **2136:** Io Volcanic Observatory research station established near Loki Patera

- **2150:** First human-piloted submarine exploration and research mission into Europa's ocean

- **2218:** Tourist excursions begin to newly expanded bases on all four large moons

and where small numbers of tourists are welcomed to learn about Ganymede and other places in the Jupiter system. Field trips to impact craters near the base are common, because scientists hope that the impacts have excavated formerly deep materials that may have been in contact with the deep ocean.

CALLISTO

Callisto, a heavily cratered world that is as large as Mercury, is the least geologically active of the four Galilean satellites. Planetary scientists believe that Callisto is less active because it orbits the farthest from Jupiter and thus does not experience the strong gravitational and tidal forces experienced by closer Io, Europa, and Ganymede. Still, there is evidence for some internal heating, and a slushy, partially liquid "mantle" layer is thought to lie below the thick icy crust. The surface itself is coated by a thick layer of powdery dust and ice created by billions of years of impact cratering. Researchers at Callisto Station also welcome limited numbers of tourists, and excursions into the field provide the opportunity for some of the best pristine-powder skiing in the outer solar system (your guides can rent you all of the proper specialized equipment). Plans are afoot to develop a true tourist-focused skiing and winter sports center on Callisto, so you might want to make plans to visit before the place gets too crowded and the powder too packed.

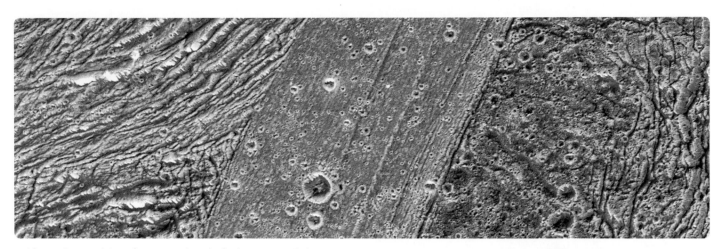

Above: Ganymede's surface is a mishmash of ridges, cratered plains, and sea ice-like plates. The area here is 55 miles (90 kilometers) wide.

Right: The surface of Callisto is heavily cratered and coated by finely powdered rocky and icy debris. The scene here is about 20 miles (32 kilometers) wide.

GETTING THERE

Getting to the moons of Jupiter presents the same challenges as getting to Jupiter itself—long trip times, high costs, limited options, and travel into dangerous environments. To truly experience these worlds, though, plan to spend a year or two of your life (at least) traveling to and exploring them all.

THINGS TO DO, PLACES TO STAY

Besides the spaceliner that brought you to the Jupiter system from Earth (or perhaps Mars), your accommodation options are limited so far to just one established base on each of the four large moons, as well as very limited numbers of rooms on a few of the submersible science research stations in Europa's ocean. Life on these moons is still heavily focused on scientific research, and so there are no 5-star resorts or spas (yet), and dining and entertainment choices are substantially more limited than in the inner solar system. Still, timing your trip to coincide with special festivals or visiting entertainment troupes can make the voyage even more rewarding. Regularly scheduled shuttles offer ferry service between all four moons, so do take the opportunity to visit more than one of these interesting and dynamic worlds.

LOCAL FLAIR

All of the bases on Jupiter's four large moons are active research stations, and many of their scientists and support staff have been living on these (or other) worlds for generations. The science and exploration culture, and the sense of being out on the frontier of solar system exploration, run deep on these stations. As such, being a curious and respectful traveler interested in the history of exploration and discovery of these worlds—or even better, interested in contributing in meaningful ways to their current exploration—will win you many friends among the local crew and staff. Opportunities abound for social interactions (the moons still receive only relatively small numbers of tourist visitors), for expanding your education, and for actively participating in cutting-edge research and exploration. Don't miss out!

TITAN

RIDE THE TIDES THROUGH THE THROAT OF KRAKEN

TITAN AND THE SPLENDORS OF SATURN

Traveling out to the Saturn system takes nearly twice as long as getting to Jupiter, but the trip is worth it. Even from Earth based telescopes, the rings of Saturn have long been recognized as the crown jewel of the solar system, and their splendor only increases as you approach and then swoop down through them to enter orbit. And Saturn's panoply of more than five dozen moons doesn't fail to delight either. The largest of them, Titan, is one of the most intriguing worlds in the entire solar system.

Opposite: Titan is the only world other than Earth in the solar system with liquid covering parts of its surface. Rather than water, however, Titan's rivers, lakes, and seas are made of liquid ethane, methane, and other hydrocarbons.

Saturn's largest moon, Titan, looms in front of the giant ringed planet in this 2012 photo from the *Cassini* orbiter mission.

Titan is larger than the planet Mercury and only slightly smaller than Jupiter's largest moon, Ganymede. What makes Titan unique, however, is that it is the only moon in the solar system with a substantial atmosphere. The surface pressure of Titan's smoggy nitrogen and methane atmosphere is 50 percent greater than the surface pressure on Earth. At that pressure and at the extremely cold temperatures of the outer solar system, many of the organic molecules in Titan's atmosphere can exist as liquids instead of gases. Indeed, early missions to Titan discovered rivers and small seas and lakes of liquid methane and ethane, as well as evidence for complex geology resulting from the erosion of the icy surface by those exotic liquids. Titan may also be somewhat of an "ocean world," like Europa, Ganymede, and Enceladus (another of Saturn's moons), because there is evidence for a liquid-water layer deep under the frozen icy surface. With wind, water, lakes, complex geology, and a dynamic interior, Titan is essentially a small planet that just happens to be in orbit around Saturn.

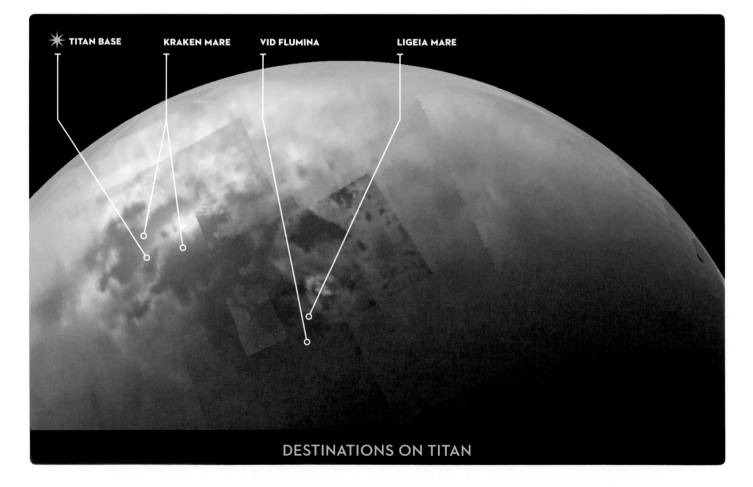

TITAN BASE   KRAKEN MARE   VID FLUMINA   LIGEIA MARE

DESTINATIONS ON TITAN

TITAN FAST FACTS

Type of Body
Planetary satellite (moon)

✳

Distance from Sun
Averages ~9.6 AU, or about 889 million mi. (1.43 billion km)

✳

Distance from Earth
Ranges from 746 million to 844 million mi. (1.20 billion to 1.36 billion km)

Travel Time from Earth
Varies from about 2 months to nearly 2 years, depending on distance

✳

Diameter
3,200 mi. (5,150 km), or about 6% larger than Mercury's diameter

✳

Highlight
Thick, smoggy atmosphere plus rivers and lakes of liquid methane and ethane!

AVERAGE TEMPERATURES

DAYTIME HIGH		NIGHTTIME LOW/SHADOWS	
°F	°C	°F	°C
-288	-178	-292	-180

TITAN AS A LABORATORY FOR THE EARLY EARTH?

A planet with a thick, nitrogen- and hydrocarbon-rich atmosphere might seem like a strange and alien place, but that is precisely what planetary scientists believe the early environment of Earth was like, 3 to 4 billion years ago. Most of early Earth's oxygen was bound up in ocean water or CO_2 dissolved in oceans, lakes, and streams, and so the atmosphere was what scientists call *reducing*—dominated by nitrogen and hydrocarbon gases emitted by volcanoes. Only after a few billion years—with the evolution of photosynthesis by simple bacteria-like life-forms—did Earth's atmosphere become as oxygen-rich as it is today. Chemistry (and any potential evolution) is much slower at the frigid temperatures of Titan, however, and so that world's reducing atmosphere has persisted for more than 4.5 billion years. Studying Titan, then, provides a unique window into understanding the early history of our own world.

BEFORE YOU GO

If you're planning a visit to Titan and other destinations in the Saturn system, there are a few things you should prepare for:

Extreme Radiation: Just like Jupiter, Saturn has a strong magnetic field that creates high-radiation challenges to people and equipment. While its radiation environment is not as intense as Jupiter's, you must still exercise caution and be sure that your safety training, and your tour operators' safety permits, are up to date.

Extreme Pressure: If you travel to the surface of Titan, you'll be walking or shuttling around in an environment with 50 percent higher pressure than Earth's atmosphere at sea level (the equivalent of diving down about 15 m—roughly 50 ft.—underwater). Your space suit and other equipment will be built for the higher pressure, but remember that your body isn't. So pay special attention to safety drills designed to help you acclimatize to the pressure and avoid decompression sickness.

CAUTION!

Beware of toxic gases. Titan's atmosphere is mostly inert nitrogen (98 percent), but the rest of the atmosphere is a noxious combination of methane, hydrogen, and other hydrocarbons such as ethane, diacetylene, methylacetylene, acetylene, and propane. Your space suit and surface accommodations will provide breathable air, of course, but the possibility of mixing Titan's hydrocarbons with oxygen creates a potentially explosive situation—literally. Be sure to follow all safety drills, and be patient and respectful of the fact that some facilities will have double or even triple airlocks to guarantee that the toxic gases stay outside.

DON'T MISS . . .

TITAN BASE AND CRUISING KRAKEN MARE

The largest sea on Titan is Kraken Mare (named after the mythological Norse sea monster), which is slightly larger than the Caspian Sea on Earth. Hundreds of bays and inlets define the shoreline of Kraken, and along the ridge near one of the most picturesque of those bays, scientists have established a permanent research station known as Titan Base. It's a modest facility, housing a team of about 50 scientists and support staff, and like other outer-solar-system bases over the past few decades, the crew have started taking on adventure tourists as a way to help defray research expenses. In addition to learning about their latest scientific results, you'll have the opportunity to join the crew on one or more research

Magical but usually quite haze-shrouded sunsets are common views from Titan Base, along the shores of Kraken Mare.

A spectacular propane waterfall along the banks of Vid Flumina, near the entry point of your amazing "white-water" rafting adventure.

cruises on their ship, the *Gaia*, which has been specially adapted for both sailing and powered travel across the liquid-hydrocarbon lakes and seas of Titan. Especially memorable is the dinner-cruise lecture series, usually led by one of the young postdocs on the crew, where you can learn about the history of Titan exploration and take the opportunity for detailed Q and A about solar system exploration. Many of the current crewmembers are also quite musically talented, and so it's not uncommon for cruises to feature nightly entertainment and dancing.

TITAN "WHITE-WATER" RIVER RAFTING

For the truly brave, you can now sign up with the Titan Base crew for a shuttle ride to a thrilling new rafting trip through the rapids of Vid Flumina, a large river that drains into the second-largest sea on Titan, Ligeia Mare. Large ice boulders eroding off the slopes of the river channel have created some spectacular class 4 and class 5 rapids on several sections of the river. While "white water" isn't quite the right term (it's not water, but liquid ethane and other hydrocarbons), the river still gets choppy, and whitecap waves are common. Your guides will include seasoned white-water-rafting specialists who have learned to understand (and respect) the daily and seasonal changes in the rapids. You'll have to go on a fairly strenuous hike (even with the lower gravity—only 15 percent of Earth's) to arrive at where the rafts are put into the river, but the sights along the way are some of the most spectacular in the solar system, reminiscent in many ways of the river-carved landscapes of the deserts in the American Southwest, or of fjord country in Norway. Where else can you hike under the mists created by a

Millions of house-size chunks of ice dance gracefully in the amazingly thin rings of Saturn. Your shuttle pilot will get you up close and personal with these boulders as you travel through the rings.

200-foot (61-meter) waterfall of propane? If you're lucky, on the way back your guide will also take you on a quick tour of the 2005 *Huygens* probe landing site, from which humanity got its first real look at this amazing world.

A LAP THROUGH SATURN'S RINGS

While you're in the neighborhood, don't miss the opportunity to see more of Saturn itself. In stunning shuttle tours through the famous rings of Saturn, pilots will deftly guide you through a maze of millions of dust-size to house-size floating chunks of ice that are confined by gravity into a thin set of rings only—amazingly—30 feet (10 meters) thick, but as wide across as 16 Earths. By matching your velocity with the ring particles, you'll be able to see them up close and hear them gently bumping against the well-protected hull of the ship. It's a scene right out of a science-fiction movie—surrounded by boulders of shimmering, spinning ice—but you can experience it for real. On the way to or from the rings,

your pilot will also take the ship on one or more dives into the upper atmosphere of Saturn itself, allowing scientists to take samples of different cloud and haze layers so that they can learn more about the second largest of our solar system's gas giant planets.

GETTING THERE

Saturn is nearly twice as far from Earth as Jupiter is, and thus travel times are correspondingly longer, ranging from about two months to two years (each way) from Earth, although shorter travel times are available with new departures from Mars and Jupiter now being offered. Also, because few tourist opportunities are yet available in the Saturn system, flights are less frequent. Seats fill up fast on the available spaceliners, so be sure to book well in advance. Once in orbit around Saturn, shuttles will dock with your spaceliner and take you down to Titan Base or Enceladus Station (see chapter 13), or on a tour of Saturn's rings and cloud tops.

HISTORY OF EXPLORING TITAN AND SATURN

- **1655:** Saturn's largest moon, Titan, discovered by Dutch astronomer Christiaan Huygens
- **1659:** Huygens discovers that Saturn has rings
- **1979:** *Pioneer 11* robotic spacecraft makes first flyby through the Saturn system
- **1980–81:** *Voyager 1* and *Voyager 2* acquire first high-resolution images of Saturn and its moons and rings
- **2004:** *Cassini* robotic spacecraft becomes first mission to orbit Saturn
- **2005:** *Huygens* probe, deployed by *Cassini*, successfully lands on Titan
- **2032:** First successful atmospheric entry probe into Saturn

- **2088:** First human mission to Saturn (launched from Mars)
- **2120:** First human landing on Titan
- **2130:** Saturn robotic "deep probe" mission samples planet's metallic hydrogen interior
- **2150:** Titan Base established along the shore of Kraken Mare
- **2210:** Routine tours of Saturn's rings become available
- **2218:** New "white-water" rafting tours offered for Titan adventure travelers

THINGS TO DO, PLACES TO STAY

Accommodation choices in the Saturn system are limited to your spaceliner and the relatively few numbers of tourist rooms at the Titan or Enceladus research stations. If you head to one of the moons, you'll be shacked up with scientists—geologists, meteorologists, astrobiologists, and others—who are actively pursuing research on these diverse outer-solar-system worlds. You'll have lots of opportunities to interact with these folks and learn about their work, as well as relax with them in their off-work hours. Dining choices are limited to cafeteria-like eateries at the bases (your spaceliner likely will offer classier fare). In addition to the opportunity to go on a sea cruise, rafting trip, or side tour of Saturn's rings, you'll have the chance to go on hikes near Titan Base with crewmembers who are doing relevant and exciting fieldwork.

LOCAL FLAIR

The turnover rate of members of the Titan Base research crew is relatively rapid, with most junior crew members coming for a year or two to work on various aspects of their theses or postdoctoral field research before returning to Earth or other bases with better laboratory and computational facilities. Many, like you, will be first-time visitors to Saturn and Titan, and will be just as awed by the experience. Still, their energy and enthusiasm for the science and adventure of exploration are infectious, and the after-hours parties are legendary. Take the time to get to know them and learn about the incredible advances they are helping to make not only in understanding Titan, but in understanding the history of Earth as well.

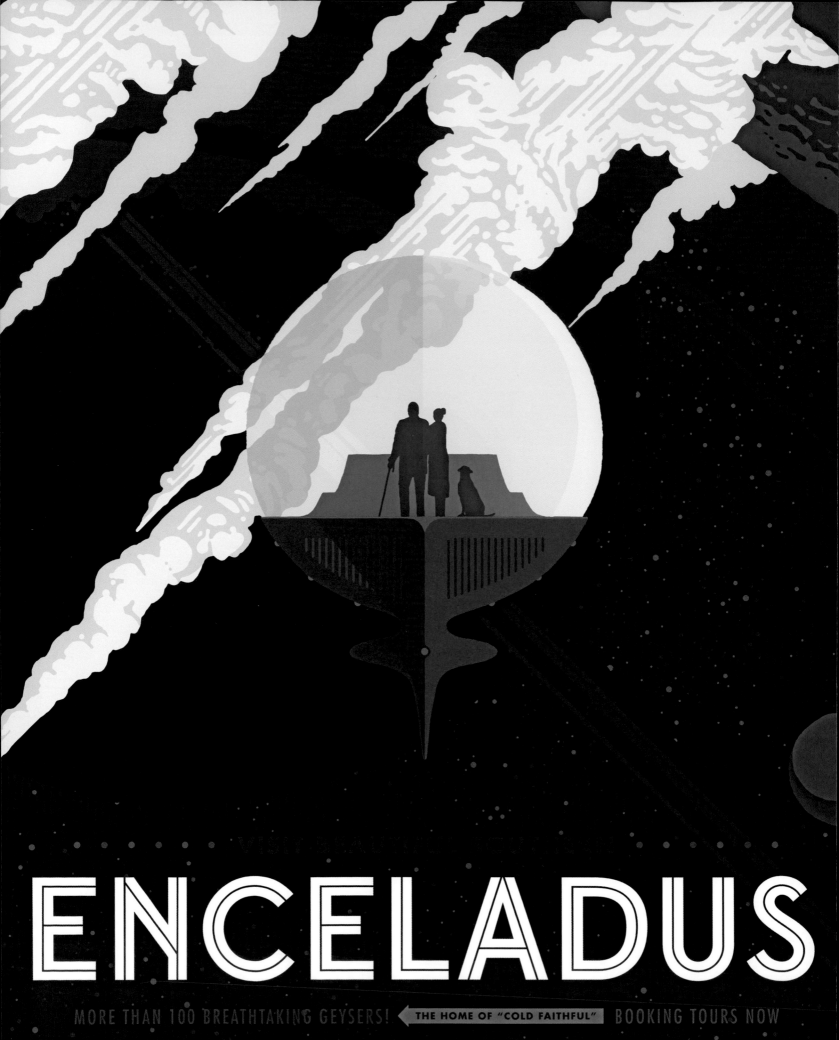

ENCELADUS AND THE ICY MOONS OF SATURN

Visiting the Saturn system provides the chance for adventure and excitement beyond just Titan and Saturn's rings. Saturn is accompanied by more than 60 other moons that offer amazing vistas, excursions, and educational opportunities for travelers. Six of them (Rhea, Iapetus, Dione, Tethys, Enceladus, and Mimas) are larger than 250 miles (400 kilometers) in diameter, and thus have significant surfaces to visit and explore. Of these, perhaps the most interesting and surprising destination is Enceladus, a moon that has emerged as one of only a few "must visit" sites for astrobiology in the solar system.

Opposite: Tour the powerful water-vapor jets that emerge periodically from the deep "tiger stripe" cracks near the south pole of Enceladus.

While Enceladus may be tiny, with a diameter of only 310 miles (500 kilometers), its bounty of internal heat has created a small ocean's worth of subsurface liquid water. That internal heat likely comes from the periodic gravitational squeezing that Enceladus gets from Saturn and other larger moons that pass nearby as they orbit. The ebb and flow of the gravitational pulls of the other worlds around Enceladus generates tidal forces that heat the interior, melting some of the icy mantle and creating a liquid-water layer under the icy crust.

One dramatic result of these tidal forces acting on Enceladus is the periodic eruption of gigantic saltwater geysers that spew from huge, warm cracks (called "tiger stripes" because of their patterning) in the icy crust near the moon's south pole. The water, which travels from the ocean up through the stripes, instantly crystallizes into an icy jetlike spray in the vacuum of space, making parts

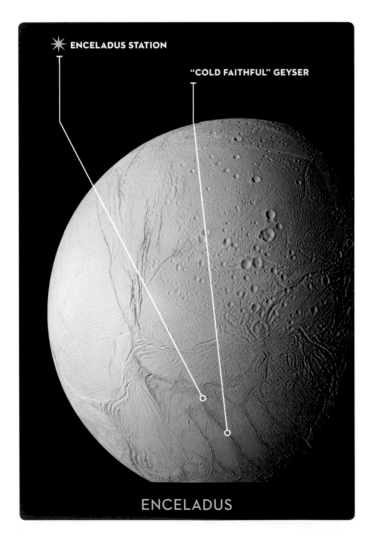

✳ ENCELADUS STATION

"COLD FAITHFUL" GEYSER

ENCELADUS

HOW DO TIDAL FORCES HEAT A PLANETARY INTERIOR?

If you've ever played racquetball or squash, you know that as the game goes on, the ball gets hotter and hotter. This is because it's getting continually stretched by the impact of the racquet. It turns out that planetary moons can act the same way. As a moon orbits, each time it passes by a neighboring moon it feels the gravitational attraction of that other moon and moves toward it. But it also feels the much larger gravity of its parent planet. The opposing forces act like the tides on an ocean, and this stretches out the shape of the intervening moon ever so slightly. If the stretching happens often enough—for example, if the inner moon orbits exactly twice for every one orbit of the outer moon (what astronomers call a resonance), then the forces can amplify over time. Repeated stretching and relaxing for billions of years can heat a moon's interior. In the case of Jupiter's innermost moon, Io, it has been heated above the melting point of rock. For other tidally heated moons like Europa and Enceladus, the heating is less extreme, but it is still enough to melt ice and create subsurface liquid-water oceans.

of Enceladus seem almost comet-like. The cloud of icy material that is ejected into space as Enceladus orbits Saturn is known as the E ring in Saturn's impressive ring system. No other small, icy moon in the solar system exhibits this kind of geologic activity. Measurements of those jets have revealed the presence of a number of moderately complex organic molecules, suggesting that the warm, wet interior of Enceladus is—surprisingly—a potentially habitable environment for life as we know it. That astrobiological potential is what drove the establishment of a permanent research station on tiny Enceladus.

ENCELADUS FAST FACTS

Type of Body
Planetary satellite (moon)

✳

Distance from Sun
Averages ~9.6 AU, or about 888
million mi. (1.43 billion km)

✳

Distance from Earth
Ranges from 746 million to 844 million
mi. (1.20 billion to 1.36 billion km)

Travel time from Earth
Varies from ~2 months to nearly 2
years, depending on distance

✳

Diameter
Only 310 mi. (500 km)

✳

Highlight
Enormous salty geysers indicate that
Enceladus is an ocean world!

AVERAGE TEMPERATURES

DAYTIME HIGH		NIGHTTIME LOW/ SHADOWS		S. POLAR "TIGER STRIPES"	
°F	°C	°F	°C	°F	°C
-330	-200	-364	-220	-310	-190

BEFORE YOU GO

If you're planning a visit to Enceladus or any of the other moons of Saturn, review the precautions for handling Saturn's high levels of radiation in the previous chapter and consider these factors as well:

Low Gravity: Like the moons of Mars and the smaller icy moons of Jupiter, most of Saturn's major icy moons have very low surface gravity, ranging from about 2.7 percent of Earth's gravity on Rhea to only 1.1 percent of Earth's gravity on Enceladus. Thus, prepare for low-g accommodations on Enceladus Station, and try to keep your low-g mobility training in mind when you are out and about on Saturn's icy moons.

Extreme Temperatures: Because most of Saturn's moons are icy, they usually have super-bright surfaces that reflect 80 percent or more of the sunlight they receive. They reflect so much sunlight, in fact, that they don't absorb much heat. Combine that with their large distances from the Sun, and you get surface temperatures that are extremely low, typically getting no warmer than about -330 degrees Fahrenheit (-200 degrees Celsius), except for in unusual places like the warm southern polar fissures of Enceladus. Thus, you'll need to train to use special thermally enhanced space suits if you want to go on excursions. Polarized sunshade visors in your helmet will help you deal with the glare from all that super-bright ice.

An artist's view of the *Cassini* spacecraft, taking a magical flight through the powerful water-vapor jets of Enceladus.

DON'T MISS . . .

ENCELADUS GEYSER HOPPING

The powerful geysers near the south pole of Enceladus deliver free "samples" of deep subsurface ocean water to the surface, supporting an active scientific-research community at Enceladus Station. Using some of the latest advances in transparent-bottomed shuttle technology, you and your family can accompany the research staff as they hop from geyser to geyser, nestling up close to each to take samples. The watery jets are powerful but (fortunately) predictable in their eruptions—earning the largest of them the name Cold Faithful—and the view of their shimmering, vaporizing ice particles sparkling in the sunlight against the backdrop of Saturn's rings is absolutely mesmerizing.

CITIZEN SCIENCE AT ENCELADUS STATION

For ecotourists who want to get actively involved in ongoing research, numerous opportunities exist at Enceladus Station. For example, you can accompany scientists on surface excursions out to the famous tiger stripes, where you can help collect samples of fresh

RHEA   IAPETUS   DIONE   TETHYS   MIMAS

geyser deposits. Or you can go on stunning hikes through enormous ice caves that course through the icy crust of Enceladus, helping to collect and document samples that go further back in geologic time. You can even volunteer to work in several of the station's laboratories, searching through decades of ice and water samples for evidence of so-called biomarkers—tiny pieces of organic molecules or other chemical signals that might help advance the search for life on Enceladus.

TOUR THE ICY MOONS

Each of the six largest icy moons of Saturn has had a different—and fascinating—geologic history, and each exhibits unique and beautiful landforms that warrant a visit. So don't just limit yourself to Titan and Enceladus: Take the opportunity to visit the enormous impact basins and central peaks of Tethys and Mimas (the latter of which evokes an image of the Death Star from the *Star Wars* movie series), hike among the inky-black dark spots of the equatorial bulge on Iapetus, cruise in a transparent-bottomed shuttle just above the deep, wispy fractures of Dione, and take a six-wheel thrill ride across the rugged, heavily cratered terrain of Rhea. Several tour companies offer these excursion opportunities. While your time on each of these exotic worlds will be relatively short, as there are no accommodations, the sheer diversity of your experiences will make the visits more than worthwhile.

Top: Five other large, icy moons—Rhea, Iapetus, Dione, Tethys, and Mimas (not to scale)—provide opportunities for additional tours and excursions in the Saturn system. Rhea is the largest at 948 miles (1,526 kilometers) across, and Mimas is the smallest at 246 miles (396 kilometers) across.

Above: Numerous excursions are offered by Enceladus Station researchers for tourists who want to experience firsthand the power of that world's giant water-vapor geysers.

HISTORY OF EXPLORING ENCELADUS AND THE ICY MOONS OF SATURN

- **1671–84:** Iapetus, Rhea, Tethys, Dione discovered by Italian astronomer Giovanni Cassini
- **1789:** Enceladus and Mimas discovered by British astronomer William Herschel
- **1980–81:** *Voyager 1* and *Voyager 2* acquire first high-resolution images of Saturn and its moons and rings
- **2004:** *Cassini* robotic spacecraft becomes first mission to orbit Saturn
- **2006:** Water-vapor plumes discovered on Enceladus from *Cassini* data

- **2052:** First robotic landing on Enceladus, followed by sample return to Earth
- **2135:** First human landing on Enceladus
- **2176:** Enceladus Station established
- **2206:** Researchers reach Enceladus subsurface ocean with robotic submersibles
- **2218:** Enceladus Station expands tourist/visitor opportunities

GETTING THERE

Visiting Enceladus or any of the other icy moons of Saturn presents the same challenges as visiting Titan or Saturn's rings: long travel times, infrequent departure and return opportunities, and generally high costs. The lower gravity and lack of atmosphere on the icy moons means that landing on those worlds is easier than landing on Titan, however. Most visitors arrive in Saturn's orbit via a major spaceliner flight and then use local shuttles to get to Enceladus Station and other icy-moon destinations. Titan and Enceladus are the only icy moons of Saturn that have bases or research stations, and so the specific landing locations and excursion options vary by tour company.

THINGS TO DO, PLACES TO STAY

Accommodations on Enceladus Station are fine but basic, as tourists have only recently started to visit a new hotel-like wing of the station. You won't find fancy spas or nightclubs, but you will get an authentic experience of what it's like to serve on a scientific-research base in the deep outer solar system. As yet there are no other bases on the icy moons, so either your spaceliner will land near the prime geologic attractions or a shuttle will ferry you from moon to moon and then back to your spaceliner or Enceladus Station. In addition to the Enceladus geyser tours and short surface excursions on the other icy moons, visitors highlight "Saturn watching" as a major activity. The ever-changing colors, patterns, and perspectives of the giant planet's clouds and magnificent rings are spellbinding.

LOCAL FLAIR

For the extreme-sports enthusiasts, a group of researchers from Enceladus Station have recently pioneered a new activity that they call geyser gliding. Essentially, a properly suited person jumps headfirst into an active geyser jet and is rapidly (sometimes violently) whisked along in the supersonic (over 1,000 miles [1,600 kilometers] per hour!) flow. Special winged space-suit flaps allow a modest amount of steering and orientation control within the jet, and a special rocket attachment can be fired to escape from the flow and return back to the surface. As yet it's a private, non-governmentally sanctioned, and potentially dangerous activity (several researchers have been critically injured), but rumors abound about thrill-seeking tourists forking over large sums of money for the chance to join the ride.

Accompany your scientist guides on spectacular hikes to the geysers
of Enceladus!

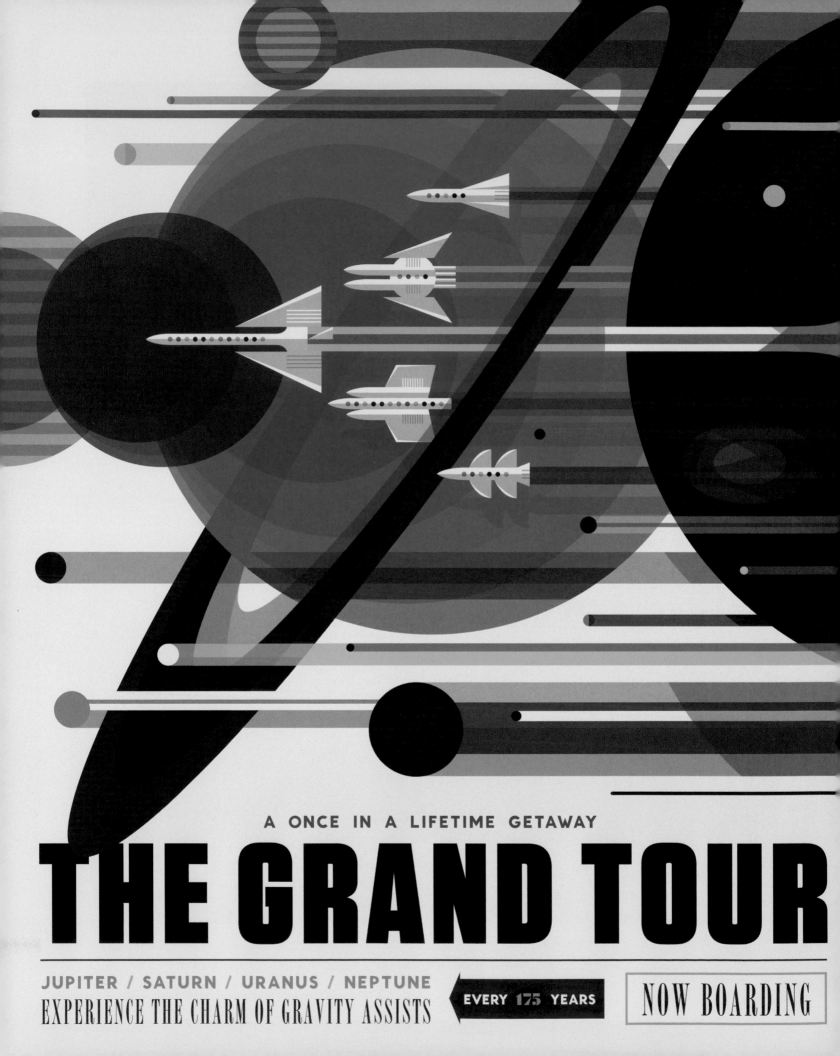

VISITING URANUS, NEPTUNE, AND PLUTO

When British astronomer Sir William Herschel discovered the planet Uranus in 1781, he instantly doubled the size of the solar system. Uranus's distance from the Sun is about 19 times farther than the Earth's and a little more than twice that of Saturn, which until then had been the most-distant known planet in the solar system. Then when Neptune was discovered in 1846 (independently by French and British astronomers), the extent of the solar system expanded again, out to 30 astronomical units (30 times the distance from the Earth to the Sun). And when Pluto, the last major planetary body of the classical telescopic age, was discovered in 1930, the scale of the solar system took yet another jump, out to 40 astronomical units. These distances are truly astronomical (40 astronomical units is almost 3.7 *billion* miles, or 6 *billion* kilometers), partly explaining why these nether regions of the outer solar system are relatively unexplored when compared to destinations closer to the Sun.

Opposite: Re-create the historic late-twentieth-century mission of the *Voyager 2* space probe by signing up for one of the new Grand Tour voyages to the outer solar system!

Happily, however, the rarity of voyages to Uranus, Neptune, Pluto, and other small worlds in the region beyond Pluto known as the Kuiper Belt also provides the chance for adventure travel *and* true science and exploration opportunities for the patient and curious solar system traveler. Indeed, the outer-solar-system research community has embraced ecotourism as a critical component for not only helping to fund new geology, atmospheric science, and other space-research expeditions to those worlds, but also staffing the research crew itself for the unavoidably long voyages. Interested in cliff jumping on the small, icy moons of Uranus? Want to see firsthand the nitrogen-ice geysers of Triton? Feel the urge to ski the powdery mountains of Pluto? Take a course in planetary mapping, exploration geophysics, or field geology on the trip there, and you'll be an expert by the time you arrive.

BEFORE YOU GO

If you're planning a visit to any of the worlds in the distant outer solar system beyond Saturn, there are a few things for which you should prepare:

🕐 **Long, Long, LONG Travel Times:** The enormous scale of our solar system was reinforced long ago when the first space probe sent to Pluto (launched in 2006) took *nine years* to get there. Travel times are generally shorter these days compared to back in the twenty-first century because of advances in propulsion technology, but trip durations for destinations beyond Saturn are still measured in years rather than the months or weeks required to reach most inner-solar-system destinations. An implication of this celestial-mechanics reality is that unless your vessel is planning to rendezvous with others on the way out or back, you won't be able to change your

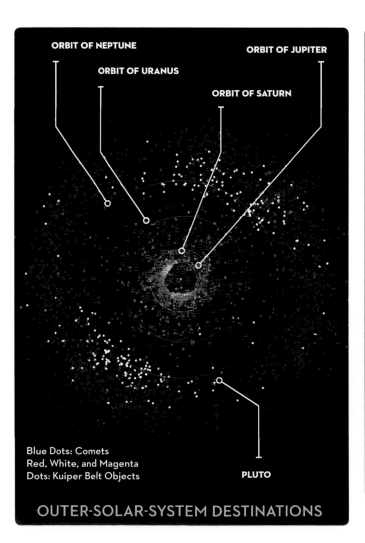

OUTER-SOLAR-SYSTEM DESTINATIONS

ORBIT OF NEPTUNE

ORBIT OF URANUS

ORBIT OF JUPITER

ORBIT OF SATURN

Blue Dots: Comets
Red, White, and Magenta
Dots: Kuiper Belt Objects

PLUTO

NEPTUNE

EARTH

MOON

PLUTO

URANUS

Above: Uranus, Neptune, and Pluto to scale, along with the Earth and Moon, for comparison.

OUTER-SOLAR-SYSTEM FAST FACTS

Types of Bodies
Uranus, Neptune: gaseous-giant planets; Pluto: small icy, rocky planet

✳

Distances from Sun
Uranus averages 19.2 AU, or about 1.8 billion mi. (2.9 billion km)
Neptune averages 30.1 AU, or about 2.8 billion mi. (4.5 billion km)
Pluto averages 39.5 AU, or about 3.7 billion mi. (5.9 billion km)

✳

Distance from Earth
Only about 1 to 2 AU less than the above distances

Travel Times from Earth
Minimum of 4 months to Uranus, 6 months to Neptune, 8 months to Pluto. Slower vessels can take up to 5 times as long, however.

✳

Diameters
Uranus: 31,765 mi. (51,120 km), 4 times the diameter of Earth
Neptune: 30,600 mi. (49,246 km), 3.9 times the diameter of Earth
Pluto: 1,479 mi (2,380 km), just under 20% the diameter of Earth

✳

Highlight
Gorgeous blue and green atmospheres on Uranus and Neptune, dozens of small icy moons, and hundreds of planetary-size bodies like Pluto extending far from the Sun.

AVERAGE TEMPERATURES

PLANET	DAYTIME HIGH		NIGHTTIME LOW	
	°F	°C	°F	°C
Uranus (cloud tops)[a]	-357	-216	-357	-216
Neptune (cloud tops)[b]	-310	-190	-360	-218
Pluto	-360	-218	-400	-240

[a] For reasons still not entirely understood, heat seems to be so uniformly distributed in the atmosphere of Uranus that the daytime and nighttime temperatures are essentially the same!

[b] Unlike Uranus, and for reasons as yet unknown, Neptune is generating prodigious internal heat, keeping it from being much colder than it would otherwise be at its great distance from the Sun.

mind and come home once you've embarked. Before you leave, you'll want to make absolutely sure that you're happy with the accommodations, dining, entertainment, and other options offered by your crew and ship.

Extreme Cold and Dark: Sunlight on Uranus is only 25 percent as bright as on Saturn (and only 0.3 percent as bright as on Earth). The Sun is even weaker on Neptune and Pluto, averaging less than 10 percent

IS PLUTO A PLANET?

It seems like a crazy question, given what we now know about this small and interesting world. However, back in the early twenty-first century, a fierce debate raged in the astronomical community about whether or not small worlds like Pluto could truly be classified as planets. One faction of scientists opted for the definition of a planet as an object that orbits the Sun, is massive enough to be spherical, and has "cleared its neighborhood" of other planets, asteroids, and the like. Since Pluto does not qualify for the third criterion (Neptune is in its "neighborhood," for example), it was dubbed a dwarf planet. Another group of scientists argued that planets should be judged on their intrinsic merits—*what* they are rather than *where* they are. By that reckoning, a body like Pluto, which is large enough to be spherical, has had a complex internal and geological evolution (including forming a core, mantle, and crust), and has an atmosphere and five moons, would be considered a full-fledged planet. In fact, if solar system bodies were to be judged on these merits, there would be more than 50 amazing planets out there to explore, including many of the large moons of the giant planets. A rose by any other name, to be sure.

of Saturn's levels, and less than 0.1 percent of Earth's levels. The result is that the Sun does not appear to be much more than a bright star in the sky, and thus the environments on and around these distant worlds are extremely cold and dark. Your ship will be well insulated and lit, but if you're planning any excursions, you'll find your space suit to be especially bulky and cumbersome because of all the necessary insulation and batteries for lighting. Fortunately, you'll have plenty of time on your trip out to practice using the equipment and planning for the local conditions.

DON'T MISS . . .

MIRANDA CLIFF JUMPING AT VERONA RUPES

The five large, icy moons of Uranus orbit around the blue-green, methane-rich gas giant planet in a bull's-eye pattern, because Uranus spins tipped on its side as it orbits the Sun. Perhaps as a result of the giant impact, or whatever other catastrophic event gave Uranus its strange tilt early in solar system history, its icy moons exhibit a wide variety of tectonic landforms such as canyons, troughs, cliffs, and ridges. Among the most spectacular are the enormous ice cliffs of Miranda, the innermost and smallest of the Uranian moons. Tiny Miranda (just under 300 miles, or 480 kilometers, across) appears to have been ripped apart and then put back together all jumbled up. In places, the reassembled pieces form steep, icy cliffs, and the tallest (called Verona Rupes) is 3–6 miles (5–10 kilometers). Join a research crew exploring the surface of this little world, where the gravity is only 0.8 percent of that on Earth, and jump feet-first off the edge of the tallest cliff in the solar system. As you fall, you'll have plenty of time to enjoy the amazing sights around you, because you'll be falling for about 8 minutes, until you land gently on the soft airbags waiting for you at the bottom.

TRITON GEYSER HIKE

Back during the 1989 *Voyager 2* flyby, one of the great surprises from the first views of Neptune's large moon Triton (which, at 1,690 miles, or 2,720 kilometers, in

Above: As your ship skims through the high upper atmosphere of Neptune, clouds, thunderstorms, and azure blue skies will surely remind you of the skies back on faraway Earth.

Below: Enormous, steep cliffs dominate parts of the surface of the small Uranian moon Miranda. Here, the cliff called Verona Rupes towers more than 3 miles (5 kilometers) above the surrounding icy plains.

diameter, is about 75 percent the size of Earth's moon) was the discovery of active geysers on the surface. With a surface so far from the Sun at a temperature just a few tens of degrees above absolute zero, no internal geologic activity was expected. And yet, there they were: enormous plumes of nitrogen gas being jetted out of surface fractures and carried aloft in the thin atmosphere by the prevailing winds. Scientists are keen to study these geysers, as they provide a way to learn about materials and conditions inside Triton. So hop into your thermally enhanced space suit and join the research team for a hike along some of the largest fractures, and enjoy the spectacular eruptions.

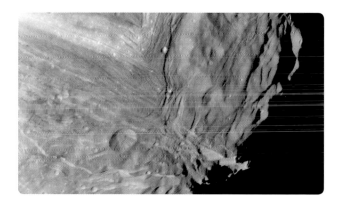

SKI PLUTO!

Even smaller than Triton and farther from the Sun, prior to the 2015 *New Horizons* flyby mission, Pluto was expected by many planetary scientists to have little or no interesting geology or surface activity. Little Pluto proved them all wrong, however, and the encounter revealed an active surface with abundant evidence of a rich geologic history shaped by tectonism, eruptions from icy volcanoes, impact cratering, and erosion. Among the most surprising features were tall mountains of water ice rising several miles above rolling plains of nitrogen, methane, and carbon monoxide ices. Early scientific expeditions to Pluto found that powdery snow coating those mountains, combined with a decent amount of gravity (about 6 percent of Earth's gravity), provided the opportunity for some of the best skiing and sledding beyond the Jupiter system. Join a research team on a surface-sampling excursion, don some specially outfitted Plutoskis, and hit the slopes!

The Sun is just another bright star in the sky as viewed from the nitrogen-ice geysers on the surface of Neptune's largest moon, Triton.

GETTING THERE

Until late in the 2100s it was only possible to get to the outer solar system beyond Saturn by booking one of the few tourist rooms provided on one of the few research vessels that headed out roughly every five years. More recently, however, several tour companies have begun offering extremely long-duration Grand Tour experiences specifically focused on adventure travel but that also include significant research and exploration components. The catch, however, is that you have to sign up for the long haul, devoting at least five years (round trip) for the flyby voyage to the distant outer solar system, and a decade or more for the new orbital opportunities at Uranus and Neptune. If you've got the time, it is a great opportunity to explore some amazing worlds out on the frontier of the solar system.

THINGS TO DO, PLACES TO STAY

Your research vessel or spaceliner will be your home away from home for at least five to ten years, so closely study the accommodations and dining options to make sure that they'll work for you. Shipboard entertainment on typical science research vessels is usually limited to the off-duty skills and interests of the crew and passengers, but previous travelers have reported outstanding musical, theatrical, and artistic performances and experiences. Spaceliner options include professional entertainers and a wider selection of dining and club choices. All of these long-duration vessels offer exercise and gymnasium-style activities, of course; even with the artificial gravity created by the ship's slow spinning motion, you'll still want to stay in shape and train for a variety of eventual low-gravity destinations.

HISTORY OF EXPLORING URANUS, NEPTUNE, AND PLUTO

- **1781:** Uranus discovered by William Herschel (first planet to be discovered by telescope)
- **1846:** Neptune discovered by Urbain Le Verrier, Johann Galle, and John Couch Adams
- **1930:** Pluto discovered by American astronomer Clyde Tombaugh
- **1986:** *Voyager 2* becomes first robotic mission to study Uranus and its moons and rings
- **1989:** *Voyager 2* becomes first robotic mission to study Neptune and its moons and rings
- **1992:** First Pluto-size Kuiper Belt Object discovered beyond the orbits of Neptune and Pluto
- **2015:** *New Horizons* becomes first robotic mission to study Pluto and its moons
- **2019:** *New Horizons* encounters Kuiper Belt Object 2014 MU69

- **2054:** *Voyager 3* robotic mission becomes first Uranus orbiter
- **2065:** *Voyager 4* robotic mission becomes first Neptune orbiter
- **2077:** *Voyager 5* mission deploys twin robotic landers on Pluto and its large moon Charon
- **2115:** *Frontier 1* becomes first human-crewed flyby, atmospheric probe, and sample-return mission to Uranus
- **2117:** *Frontier 1* human mission performs Neptune flyby, atmospheric probe, sample return
- **2160:** First human landings on Pluto and Charon
- **2190:** Routine twice-per-decade research cruises begin to the Uranus and Neptune systems
- **2218:** Grand Tour tourist cruises begin, offering decade-long flybys of the Jupiter, Saturn, Uranus, Neptune, and Pluto systems

LOCAL FLAIR

Because of the long duration of the outer-solar system travel opportunities beyond Saturn that are currently available, passenger manifests on the voyages have tended to be dominated by retirees and others who acknowledge that the trip might ultimately end up being one way. Passengers (or crew) who pass away on the long trips to or from these distant worlds are afforded full "burial at sea" rites, following the long tradition of long-voyage sailors back on Earth. Rather than morbid occasions, however, such occurrences are often requested by spouses or families (or travelers themselves, in advance health-care directives) to be cause for celebration of a life well traveled and well lived.

Brighter Pluto and its smaller, darker moon Charon were among the main targets of the 2015 *New Horizons* flyby mission.

EXTRASOLAR VOYAGES: TRAPPIST-1 AND BEYOND!

For most travelers, our solar system holds too many wonders to visit in a lifetime. But if the call of the frontier beckons to you, you're in luck. Since the late twentieth century, astronomers have been discovering exoplanets—planets orbiting stars other than our Sun—and believe there are probably millions of other relatively nearby worlds out there waiting to be explored. Some are Earthlike and could conceivably be hospitable to life as we know it, and others are gas giants, ice giants, or other kinds of exotic planets (and moons) that we can't even imagine.

Opposite: Imagine the incredible tour you could take from one Earthlike world to six others on your voyage of a lifetime to the extrasolar planetary system around the star called TRAPPIST-1.

Among the most interesting and exciting of these distant worlds is the extrasolar planetary system discovered around a star known as TRAPPIST-1. The star itself is a dim and fairly unremarkable red dwarf almost 40 light-years (about 235 trillion miles, or about 378 trillion kilometers) from our solar system. But back in 2015, astronomers detected three Earth-size planets orbiting that star, and that number went up to seven terrestrial planets in 2017. During much of the rest of the twenty-first century, astronomers were able to characterize these worlds enough to know that several have atmospheres that possibly contain life-indicating gases such as oxygen and ozone. Still, telescopes on Earth and in nearby space were not powerful enough to resolve features on these worlds, or to truly determine if life existed on any of them.

An international consortium of space agencies launched the first robotic probe to the TRAPPIST-1 system, *Odysseus*, back in 2080. Accelerated to nearly

Above: For most of the time since their discovery, the seven Earth-size planets around the red dwarf star TRAPPIST-1 were only known as faint points of light. What are these worlds really like?

Below: The seven Earthlike planets discovered in the TRAPPIST-1 system (top) compared in size to the four terrestrial planets in our solar system (bottom).

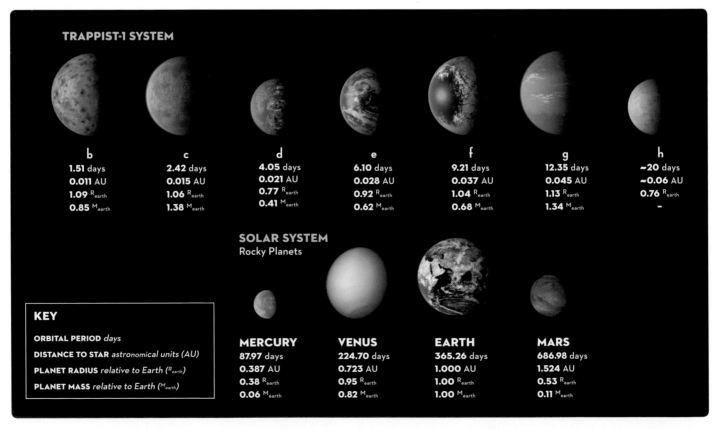

TRAPPIST-1 SYSTEM

b	c	d	e	f	g	h
1.51 days	2.42 days	4.05 days	6.10 days	9.21 days	12.35 days	~20 days
0.011 AU	0.015 AU	0.021 AU	0.028 AU	0.037 AU	0.045 AU	~0.06 AU
1.09 R_{earth}	1.06 R_{earth}	0.77 R_{earth}	0.92 R_{earth}	1.04 R_{earth}	1.13 R_{earth}	0.76 R_{earth}
0.85 M_{earth}	1.38 M_{earth}	0.41 M_{earth}	0.62 M_{earth}	0.68 M_{earth}	1.34 M_{earth}	–

SOLAR SYSTEM
Rocky Planets

KEY

ORBITAL PERIOD *days*
DISTANCE TO STAR *astronomical units (AU)*
PLANET RADIUS *relative to Earth* (R_{earth})
PLANET MASS *relative to Earth* (M_{earth})

MERCURY	VENUS	EARTH	MARS
87.97 days	224.70 days	365.26 days	686.98 days
0.387 AU	0.723 AU	1.000 AU	1.524 AU
0.38 R_{earth}	0.95 R_{earth}	1.00 R_{earth}	0.53 R_{earth}
0.06 M_{earth}	0.82 M_{earth}	1.00 M_{earth}	0.11 M_{earth}

TRAPPIST-1 SYSTEM FAST FACTS

Types of Bodies
Terrestrial (rocky, Earthlike) planets, seven total, orbiting a red dwarf star about the size of Jupiter but much heavier—about 8% the mass of the Sun

✳

Distance from TRAPPIST-1
Planets range from 1.1 to 5.5 million mi. (1.7 to 8.9 million km) from their host star, and take only between 1.5 and 19 Earth days to orbit their host star. If all seven planets moved to our solar system, they would fit well inside the orbit of Mercury.

✳

Distance from Earth
Almost 40 light-years (235 trillion mi., or ~378 trillion km)

Travel Time from Earth
Estimated to be ~80-~150 years, depending on the propulsion technologies used

✳

Diameter
The seven planets range in size from 23% smaller to 13% larger than Earth.

✳

Highlight
First discovery of extraterrestrial life of any kind beyond our solar system, on multiple habitable Earthlike planets!

AVERAGE TEMPERATURES

TRAPPIST-1 Planet	DISTANCE FROM HOST STAR (IN AU, WHERE 1 AU = 93 MILLION MI., OR 150 MILLION KM)	AVERAGE DAYTIME TEMPERATURE	
		°F	°C
b	0.011	261	127
c	0.015	156	69
d	0.021	59	15
e	0.028	-8	-22
f	0.037	-65	-54
g	0.045	-101	-74
h	0.060	-157	-105

half the speed of light using nuclear and solar-sail technology, the probe was able to get to TRAPPIST-1 by 2170, though it took another 40 years for the radio signals (traveling at the speed of light) containing the probe's images and other data to finally arrive back at Earth. Though now common knowledge, the results received in 2210 were spectacular: evidence of thick, potentially breathable atmospheres; complex geology; and plant life on at least three of the seven Earth-size worlds. But our views of those planets are still fuzzy by today's solar system–exploration standards, and it's still not clear if there is any intelligent life on those worlds. The probe detected no unambiguous extraterrestrial radio signals, although it spent only about an hour flying rapidly through and past the TRAPPIST-1 system.

The implications of the discovery of extraterrestrial life (plants and likely microbes, at least) in the TRAPPIST-1 system have awakened a desire among many scientists and explorers to mount a human voyage to this nearby solar system. And that's just what's happening, as an international consortium has announced plans to launch, within a decade, *Earth 2.0*, a modified spaceliner cruise ship that will be able to carry more than 1,000 people to the TRAPPIST-1 system. So, here's your big chance to get away from it all (literally), and take not just the voyage of a lifetime, but probably of your kids' and grandkids' lifetimes as well (the trip will take generations), as part of a bold effort to extend the reach of humanity out to the stars. It's the ultimate getaway vacation!

BEFORE YOU GO

If you're planning to apply for a spot on the spaceliner *Earth 2.0* (humanity's first planned multigenerational "starship" vessel designed for interstellar travel), which is currently being constructed to make the historic journey

The interior design of *Earth 2.0* will mimic the layout of a colony or small town, with neighborhoods, recreational facilities, and agricultural spaces lining the ring of the ship's spinning Earth-gravity section.

Initial conceptual designs of the *Earth 2.0* starship are based on centuries of NASA research on potential interstellar spacecraft systems design.

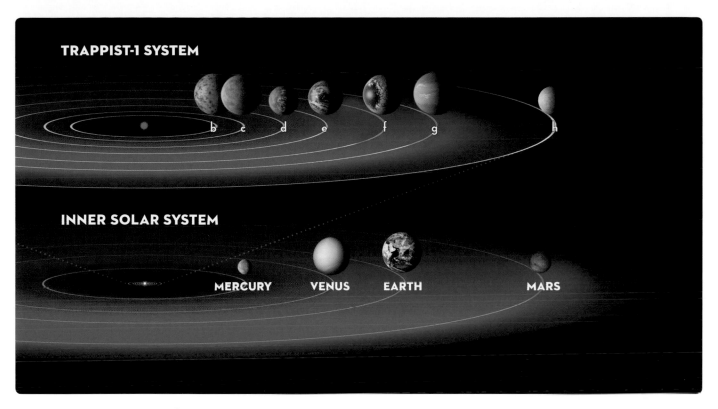

If they were in our solar system, all seven TRAPPIST-1 planets would orbit much closer to the Sun than Mercury does.

to TRAPPIST-1, there are some important things you should prepare for:

Long, Long, LONG travel times: Even at the maximum possible speed using the latest propulsion technology, mission planners still figure that *Earth* 2.0 will take at least 80 years to get to the TRAPPIST-1 system, and perhaps almost double that if the technology doesn't pan out on the scale needed for the ship. That means that unless you're a child when the mission launches, you probably won't get to visit those exoplanets in your lifetime; indeed, the ship might not even get there in your *children's* lifetimes. So you'll have to take a multigenerational mindset, and indeed, preferences for many slots on the voyage are already being given to couples with young children or firm plans to have children. If you're an older traveler, you will still be welcome; however, make sure your estate planning is settled back on Earth, and prepare to make *Earth* 2.0 your new home—forever.

Battling Boredom: A permanent move to a cruise ship-like colony in space may sound exciting, but in the absence of concrete details about who else is coming and exactly what kinds of recreational, entertainment, and dining options will be available on the ship, you should apply only if you're prepared to keep yourself busy and productive. Indeed, applications from craftsmen and artisans of all kinds—as well as artists, writers, musicians, teachers, and athletes with specific skill sets—are likely to be given preference. Think carefully about how you'll spend your time, because once you head out for the TRAPPIST-1 system, there's likely to be no turning back.

Uncertain Destinations: Even though the *Odysseus* flyby probe returned stunning images and other data about the TRAPPIST-1 planets from its quick pass through the system in 2170, there is still much that we do not know about these worlds. Thus there is

Perhaps the most spectacular image returned from the *Odysseus* probe to TRAPPIST-1 shows what looks like an inviting and hospitable surface on TRAPPIST-1d. Will it be possible for humans to live there?

significant risk that when *Earth 2.0* finally arrives in that system sometime next century, the detailed reality of those worlds may not meet expectations. What if there are as-yet undetected toxic compounds in some of the atmospheres of those planets? What if the geology is too active and violent? What if there are hostile life-forms (intelligent or not) already living there? Everyone going (and their offspring) must accept that there is a significant risk of disappointment, and mission designers are planning for the worst-case possibility that the ship itself may need to serve as a permanent home for humans in the TRAPPIST-1 system. Still, most people who've already been selected share a sense of optimism and excitement

that at least one of the worlds around TRAPPIST-1 could eventually turn into a true Earth 2.0.

GETTING THERE

A voyage to the TRAPPIST-1 system will be the last vacation of your life, so you better make it count. The enormous multigenerational starship *Earth 2.0* is being designed to emulate some aspects of long-duration spaceliners and the ocean cruise liners on Earth and Titan, but with additional facilities and amenities that acknowledge the unprecedented length of the voyage. These include a gentle spin that will give the ship a

Data from the *Odysseus* flyby probe suggest that the sunset views from the fifth of TRAPPIST-1's seven Earthlike worlds could be similar to this.

level of artificial gravity roughly equivalent to Earth's gravity, more-traditional housing and urban spaces that can simulate the look and feel of neighborhoods and being outdoors on a planetary surface, and expansive agricultural areas where crops can be grown to sustain the colony. Since it will not be possible to bring everything that is needed for a century-long (or more) voyage into deep space, and opportunities to acquire new natural resources will be limited to a rogue passing asteroid or comet now and then, at best, the level of recycling and reprocessing of air, water, and waste will need to be much more comprehensive than ever before. Indeed, many of the innovations in sustainability needed for *Earth 2.0* will be extremely beneficial to the inhabitants of Earth itself.

THINGS TO DO, PLACES TO STAY

The starship will be your new home, and the only home that your children and perhaps even your grandchildren will know for most of their lives—and potentially everyone's home forever depending on what the TRAPPIST-1 worlds are really like. Thus, things to do on this epic voyage will depend on your own work, hobbies, and interests, as well as those of your fellow passengers and crew. Mission designers are creating spaces on the ship where businesses, clubs, and recreational facilities will be able to arise and change organically and communally, depending on people's interests and needs. In a sense, life on the ship will emulate life in a newly formed colony or base back on Earth or elsewhere in the solar system, but without the freedom to leave and visit somewhere else for a long, long time.

Yet another potential vista from one of TRAPPIST-1's seven Earthlike worlds.

HISTORY OF EXOPLANET DISCOVERIES, INCLUDING TRAPPIST-1

1600: Italian astronomer and philosopher Giordano Bruno burned at the stake for suggesting the possibility of inhabited worlds around other stars

1992: First planets discovered outside our solar system, orbiting a pulsar (the high-density, rapidly spinning starlike remains of a supernova explosion) some 2,300 light-years away

1995: First extrasolar planet (51 Pegasi b) discovered around a normal Sunlike star

2009: *Kepler* space telescope begins detecting Earth-size planets around nearby stars

2018: *James Webb Space Telescope* launched; begins to characterize exoplanet atmospheres

2050: Number of known exoplanets within ~100 light-years of the Sun exceeds 100,000

2067: New *Planet Hunter* space telescope data provide evidence of water vapor and high levels of oxygen and ozone in three of the TRAPPIST-1 planets

2080: *Odysseus* robotic probe launched to the TRAPPIST-1 system

2170: *Odysseus* probe flies through the TRAPPIST-1 system at high speed (~1 hour flyby)

2210: *Odysseus* data received back on Earth; evidence for life discovered on three planets!

2218: Construction begins on *Earth 2.0*, a multigenerational starship designed to carry 1,000 people (and their offspring) to the TRAPPIST-1 system by sometime in the 2300s

LOCAL FLAIR

While the details are still being finalized, it is likely that nuclear propulsion will be used to accelerate the ship to nearly half the speed of light for the first half of the trip, and then to decelerate on the second half for eventual capture around the TRAPPIST-1 system's central star. Thus, one advantage that you and your fellow passengers will have is that, as explained by Einstein's theory of special relativity, you will age about 15 percent more slowly than everyone you leave behind. That is, if the trip ends up taking 80 years as measured back on Earth, it will only seem like 68 years to those on the starship. For future interstellar voyages that can someday travel at speeds even closer to the speed of light, the amount of time appearing to pass for the locals on the ship will be an even smaller fraction of the time elapsed back on Earth, helping to make interstellar travel opportunities for humans even more practical.

Perhaps one day a starship voyage will be mounted to the first extrasolar planet ever discovered (back in 1995)—a gas giant planet about half the mass of Jupiter known as 51 Pegasi b, located about 50 light-years from our solar system.

NOTES AND FURTHER READING

The facts, statistics, and other information about the exciting travel destinations described in this guide are based on results gleaned from the long history of telescopic and spacecraft exploration of our solar system (and beyond). The additional resources listed below can provide you with even more in-depth details about what it would be like to visit these worlds.

INTRODUCTION

Take an online tour of the solar system at the Nine Planets web site: nineplanets.org.

Curious about astronomy in general? Visit the Ask an Astronomer web site at curious.astro.cornell.edu.

Need to check the facts on the latest astronomy news? Check out Phil Plait's *Bad Astronomy* blog: www.syfy.com/tags/bad-astronomy.

To learn more about the NASA/JPL *Visions of the Future* space exploration posters featured in this book, visit www.jpl.nasa.gov/visions-of-the-future/about.php.

CHAPTER 1: WEEKENDING ON THE MOON

For more technical details on place-to-place and hour-by-hour temperature variations on the Moon, see the article www.space.com/18175-moon-temperature.html as well as the website for the NASA instrument called Diviner (www.diviner.ucla.edu) that has measured the temperature of the Moon's surface from orbit.

For a full list of and details on all of the early US *Surveyor* and USSR *Luna* robotic missions, see en.wikipedia.org/wiki/Surveyor_program and en.wikipedia.org/wiki/Luna_programme.

For more details about efforts to preserve the historic nature of the first human-landing sites on the Moon, see the following NASA document: www.nasa.gov/sites/default/files/617743main_NASA-USG_LUNAR_HISTORIC_SITES_RevA-508.pdf.

For more details on the physics behind *Apollo 14* astronaut Alan Shepard's golf experience on the Moon, check out astrophysicist Ethan Siegel's blog post at scienceblogs.com/startswithabang/2010/10/02/could-you-really-hit-a-golf-ba.

CHAPTER 2: HEATING UP ON VENUS

For more details on the early history of robotic exploration of Venus, see the website *Venera: The Soviet Exploration of Venus* by Don Mitchell at mentallandscape.com/V_Venus.htm.

For a more general compilation of *all* Venus missions to date, see The National Space Science Data Center's chronology of Venus exploration, at nssdc.gsfc.nasa.gov/planetary/chronology_venus.html.

A wonderful and readable personal account of Venus exploration can be found in planetary scientist David Grinspoon's book *Venus Revealed: A New Look Below the Clouds of Our Mysterious Twin Planet* (Basic Books, 1998).

CHAPTER 3: ZIPPING AROUND ON MERCURY

Planetary scientist Robert Strom has written an engaging book on the history of observations of the first planet from the Sun in *Mercury: The Elusive Planet* (Cambridge University Press, 1987).

NASA's *MESSENGER* Mercury orbiter mission (the acronym stands for MErcury Surface, Space ENvironment, GEochemistry, and Ranging) maintains a detailed account of the history of the mission and its results at messenger.jhuapl.edu.

CHAPTER 4: SUMMER VACATION ON MARS!

The Planetary Society, the world's largest public space-advocacy organization, hosts information about the latest Mars science and missions to Mars at www.planetary.org/explore/space-topics/space-missions/missions-to-mars.html.

A detailed chronology of the exploration of Mars by all of the world's space agencies can be found at nssdc.gsfc.nasa.gov/planetary/chronology_mars.html.

For photos and stories about the first long-range Mars rovers, *Spirit* and *Opportunity*, check out my books *Postcards from Mars: The First Photographer on the Red Planet* (Dutton, 2006) and *Mars 3-D: A Rover's-Eye View of the Red Planet* (Sterling, 2008).

Mars: The Pristine Beauty of the Red Planet (University of Arizona Press, 2017) by planetary scientist Alfred McEwen and colleagues showcases some of the most spectacular orbital views of the surface of the Red Planet ever acquired.

CHAPTER 5: FIELD TRIP TO PHOBOS

Astronomer Asaph Hall's discovery of the moons of Mars is described in detail in "The Discovery of the Satellites of Mars" (*Monthly Notices of the Royal Astronomical Society*, vol. 38, pp. 205–9, 1878), which is available online at tinyurl.com/9cy46pc.

CHAPTER 6: SMOOTH JAZZ ON DEIMOS

Detailed mosaics and maps of Deimos, Phobos, and many other irregularly shaped small worlds are available from planetary scientist Phil Stooke's NASA Planetary Data System website at pds.nasa.gov/ds-view/pds/viewDataset.jsp?dsid=MULTI-SA-MULTI-6-STOOKEMAPS-V2.0.

CHAPTER 7: CLOSE ENCOUNTERS WITH NEAR-EARTH ASTEROIDS

Astronomer Jacqueline Mitton and I co-edited a book describing the mission and results from the NASA *NEAR Shoemaker* spacecraft's voyage to the *Near-Earth Asteroid Eros: Asteroid Rendezvous:* NEAR Shoemaker's *Adventures at Eros* (Cambridge University Press, 2002).

Three different feature-length theatrical films were made in Japan about the dramatic story of *Hayabusa*, the first spacecraft to visit the near-Earth asteroid Itokawa. See, for example, the Internet Movie Database entry for *Hayabusa: The Long Voyage Home*, at www.imdb.com/title/tt1825130.

The *Wikipedia* page on Apophis (en.wikipedia.org/wiki/99942_Apophis) provides an enormous amount of information and detail about previous and future close passes of this Potentially Hazardous Object to Earth.

CHAPTER 8: GET SOME SUN!

A great general introduction to the way that stars like the Sun work can be found in astronomer James Kaler's book *Stars* (Scientific American Library, 1992).

NASA and the European Space Agency jointly operate the Solar and Heliospheric Observatory (SOHO) mission that monitors the Sun from space. Details and spectacular photos, movies, and other data can be found here: sohowww.nascom.nasa.gov.

Vladimir Bodurov has developed a fun and educational online application that lets you fly around and view the stars in the Sun's neighborhood: www.bodurov.com/NearestStars.

CHAPTER 9: TOURING THE MAIN ASTEROID BELT

For more details about the geology and chemistry of Vesta, see my articles "Dawn's Early Light: A Vesta Fiesta!" and "Protoplanet Closeup" in the November 2011 and September 2012 issues of *Sky & Telescope* magazine.

For the latest images and other results from the NASA *Dawn* mission that orbited both Vesta and Ceres, see dawn.jpl.nasa.gov.

Arizona State University's School of Earth and Space Exploration runs the *Psyche* mission, the first spacecraft to visit a metallic asteroid. Find out more details here: sese.asu.edu/research/psyche.

CHAPTER 10: EXPLORE JUPITER AND THE GREAT RED SPOT

To learn much more about Jupiter and its rings, moons, and magnetic field, download a free 2007 NASA Special Publication called "Mission to Jupiter: A History of the Galileo Project," (by Michael Meltzer) from here: tinyurl.com/3gfnqge.

For photos and fun insider stories about the 1994 impact of comet Shoemaker-Levy 9 into the atmosphere of Jupiter, see *The Great Comet Crash: The Collision of Comet Shoemaker-Levy 9 and Jupiter*, by John Spencer and Jacqueline Mitton (Cambridge University Press, 1995).

Planetary atmospheric scientist Andy Ingersoll's article "Atmospheres of the Giant Planets" provides lots more detail about the past, present, and likely future history of the Great Red Spot. The article appears as chapter 15 in *The New Solar System* (J. Kelly Beatty, Carolyn Collins Petersen, and Andrew Chaikin, editors; Sky Publishing, 1999).

CHAPTER 11: VISITING EUROPA AND THE MOONS OF JUPITER

Nearly 200 telescopic and spacecraft views of Europa are featured on the NASA Planetary Photojournal's Europa search page, at photojournal.jpl.nasa.gov/feature/europa.

Planetary scientist Rick Greenberg's popular-science book *Europa—The Ocean Moon: Search for an Alien Biosphere* (Springer, 2005) provides a great history of how and why we know that there is a deep ocean residing under that moon's thin, icy crust.

Ganymede is the primary target of the European Space Agency's *Jupiter Icy Moon Explorer* or JUICE mission. Find out more at sci.esa.int/juice.

For a beautifully illustrated and highly informative book about the volcanoes of Io and other extraterrestrial worlds, check out *Alien Volcanoes* (Johns Hopkins University Press, 2008) by planetary scientist Rosaly Lopes and space artist Michael Carroll.

Planetary scientist Paul Schenk's *3D House of Satellites* blog (at stereomoons.blogspot.com/2009/10/galileo-4-moons-at-400-years.html) provides all kinds of photos, maps, and stories about what Callisto and Jupiter's other large moons are like up close.

CHAPTER 12: TITAN AND THE SPLENDORS OF SATURN

NASA's *Cassini* spacecraft provided our first detailed, orbital views of Saturn and its rings, moons, and magnetic field. Check out spectacular images and other data from the mission at saturn.jpl.nasa.gov.

Photos and other information about Saturn's rings, as well as the ring systems around the other giant planets, can be found on the NASA Planetary Data System's "Ring-Moon Systems Node" at pds-rings.seti.org/saturn.

For an educational and highly readable account of the importance of studying Titan to understanding the history of our own home planet, see the article "The Moon That Would Be a Planet" by planetary scientists Ralph Lorenz and Christophe Sotin in the March 2010 issue of *Scientific American* magazine.

CHAPTER 13: ENCELADUS AND THE ICY MOONS OF SATURN

Facts, photos, and details on the discovery of water-vapor plumes on Enceladus can be found on that moon's extensive *Wikipedia* page: en.wikipedia.org/wiki/Enceladus.

The home page of the NASA *Cassini* mission's Imaging Team at www.ciclops.org contains spectacular images of all the other sizeable, icy moons of Saturn, as well as images of the rings and many smaller moons.

CHAPTER 14: VISITING URANUS, NEPTUNE, AND PLUTO

Some popular-science and historical accounts of the *Voyager* mission and the first Grand Tour of the outer solar system can be found in historian Stephen Pyne's book *Voyager: Exploration, Space, and the Third Great Age of Discovery* (Viking, 2010), and my own book *The Interstellar Age: Inside the Forty-Year Voyager Mission* (Dutton, 2015).

Fascinating historical accounts of the discoveries of Uranus, Neptune, and Pluto can be found in *The Georgian Star: How William and Caroline Herschel Revolutionized Our Understanding of the Cosmos* (Michael Lemonick; W. W. Norton, 2009), *The Planet Neptune: An Historical Survey Before Voyager* (Sir Patrick Moore; Wiley, 1996), and "The Search for the Ninth Planet, Pluto" (Clyde Tombaugh; 1946—the article is online at tinyurl.com/8redhe8), respectively.

The website for the *New Horizons* mission at pluto.jhuapl.edu provides photos, stories, and other data from that spacecraft's historic first encounter with Pluto in 2015.

CHAPTER 15: EXTRASOLAR VOYAGES: TRAPPIST-1 AND BEYOND!

The original NASA press release revealing the existence of seven planets orbiting around the star TRAPPIST-1 and providing additional background and details can be found at www.nasa.gov/press-release/nasa-telescope-reveals-largest-batch-of-earth-size-habitable-zone-planets-around.

Updated census counts of the currently known extrasolar planetary systems, including their physical and orbital characteristics, are maintained online from the NASA *Kepler* mission (www.nasa.gov/kepler/discoveries) and the *Extrasolar Planets Encyclopaedia*'s "Interactive Extra-Solar Planets Catalog" (exoplanet.eu/catalog).

Space artist Ron Miller's book *Extrasolar Planets: Worlds Beyond* (Twenty-First Century Books, 2002) provides a great introduction to the discoveries of planets beyond our solar system, as well as beautifully illustrated speculations about what those worlds could be like.

ACKNOWLEDGMENTS

I am grateful to the many space artists and visual strategists out there who have created the fanciful and inspirational illustrations and posters that appear in this travel guide. The art produced by Ron Miller (www.black-cat-studios.com/), Tyler Nordgren (www.tylernordgren.com/) and the artists from the NASA/JPL "Visions of the Future" project (www.jpl.nasa.gov/visions-of-the-future/about.php) has been particularly inspirational to this book. I also want to thank Meredith Hale and Linda Liang from Sterling Publishing as well as Katherine Furman and Ashley Prine from Tandem Books for their incredible editorial and artistic help, Michael Bourret from Dystel, Goderich, & Bourret for his never-ending enthusiasm, and my dear friend Jordana Blacksberg for being my constant companion and muse—you inspire me to dream of such a wonderful future!

INDEX

Note: Page numbers in **bold** indicate/include captions or maps.

IMAGE CREDITS

Front cover (jacket) photograph: JDawnInk/
iStock
Back cover(jacket) photographs: NASA/
JPL-Caltech/University of Wisconsin (back-
ground); insets clockwise from top left:
NASA/GSFC/Arizona State University; ESA/
DLR/HRSC Team; NASA; Ron Miller (2);
NASA/JPL/Bjorn Jo
Endpapers a–d (front, left to right): NASA/JPL-
Caltech; ESO/M. Kornmesser
Endpapers e–h (back, left to right): NASA/KSC;
NASA/JPL-Caltech
AKG: © Universal Images Group/Sovfoto: 20 left
Alamy: Science History Images: 82 right; Stock-
trek Images, Inc.: 52
Jordana Blacksberg: jacket (author)
© **Don Dixon:** 31
ESA: DLR/HRSC Team: 40 top, back cover;
Hubble: v, 92, 95 right (plume), 97 bottom
© **Fabled Creative/www.fabledcreative.com:**
Maxwell Montes: 18 right
© **Indelible Ink Workshop:** xiv, 1, 72, 73
Courtesy of ISAS/JAXA: 69
iStock: © Devrimb: 60 (rider); © JDawnInk: cover;
© layritten: 60 (space); © StephanHoerold:
68 top

M. Kornmesser/ESO: 138, 140, endpapers
© **Lynx Art Collection:** 24, 25
© **Ron Miller:** 55, 63, 97 top, 105 top, 112, 113, 114,
123, 129 top, 130, back cover (x2)
© **Don P. Mitchell:** 19 bottom
NASA: iv, v, 5, 10, 11 bottom, 22, 23, 39, 42, 44
right, 61 bottom, 70, 77, 79, 99, 107 top, 116, 117,
118, 120, 121, 129 bottom, 131, 134, 137, 139, back
cover; DLR/Cornell University/Phil Stooke:
50; ESA/Hubble Space Telescope: 95 left;
ESA/J. Nichols (University of Leicester): 96;
Gordon Legg/Lunar and Planetary Institute:
40 bottom; GSFC/LROC/Arizona State
University: 2, 6, 8, back cover; JHU/APL:
30, 66; Johns Hopkins University Applied
Physics Laboratory/Carnegie Institution of
Washington: 29 bottom; JPL: v, 104 top; JPL/
ASU: 107 bottom; JPL/Bjorn Jonsson: back
cover; JPL-Caltech: xi, xii, 14, 15, 34, 35, 45,
56, 57, 80, 81, 90, 91, 124, 100, 101, 108, 109,
125, 132, 133, 136, 141, endpapers (x2); JPL-
Caltech/ASU: 43; JPL-Caltech/MPS/DLR/
IDA: 82 left; JPL-Caltech/SETI Institute: 102,
104 bottom; JPL-Caltech/Space Science
Institute: 110 top; JPL-Caltech/UCLA/MPS/
DLR/IDA: 84 top, 85, 89; JPL-Caltech/Uni-
versity of Arizona: iv, 53 left and top, 58, 60

top, 61 left and right; JPL-Caltech/University
of Arizona/University of Idaho: v, 110 bottom,
111; JPL-Caltech/University of Wisconsin:
throughout (milky way); Lunar and Planetary
Institute: 9; KSC: vii, viii, ix, endpapers;
NSSDC Photo Gallery/ESA Images: 20 right;
SWRL/MSSS/Jason Major: 94
Tyler Nordgren: 105 bottom, 121 bottom
© **Jay Pasachoff/Williams College:** 76
Peter Rubin/Iron Rooster Studios/ASU: v, 82, 86
Science Source: David P. Anderson, Southern
Methodist University/NASA: 19 top
Shutterstock.com: structuresxx: throughout
(Milky Way)
**Ted Stryk/Russian Academy of Sciences/Pho-
bos-2 mission:** 53 right
USGS Astrogeology Sciences: 16, 31
Courtesy of Wikimedia Foundation: Goodvint:
5 bottom; JHUAPLC/CIW/Jason Perry: iv, 28
middle, 29 top, 30; kelvinsong: 74; NASA: iv,
v, 18 left, 20 top, 28 left, 44 left, 68, 126 right;
NASA/Jet Propulsion Lab/USGS: 36, 128;
NASA/JPL-Caltech: 64; Gregory H. Revera:
iv, 6 top, 7 bottom, 8 bottom, 11 top, 13, 28
right; Space X: 41, 48, 49
© **Gareth Williams/IAU Minor Planet Center:**
84 bottom

VENUS The surface of Venus is a hellish environment because of the planet's thick carbon dioxide atmosphere. The atmospheric pressure at the surface is more than 90 times the surface pressure on Earth, and the temperature can exceed 870 degrees Fahrenheit (465 degrees Celsius)—hot enough to melt lead! Thus, many futuristic concepts of tourism on Venus, like the one depicted in this poster, focus on the possibility of hotels and resorts that could float high up in the atmosphere, where the temperatures and pressures are much more Earthlike.

Credit: Courtesy NASA/JPL-Caltech

MARS Nearly 50 robotic space missions to explore Mars have been attempted since the start of the space age. Even though only about half of them have been successful, Mars is still one of the most-studied places in the solar system beyond Earth. For future tourists interested in the history of space exploration, opportunities will abound to visit some amazing historic sites on Mars: first robotic landing, first rover, places from which the first samples were returned to Earth, and, ultimately, the first human landing site.

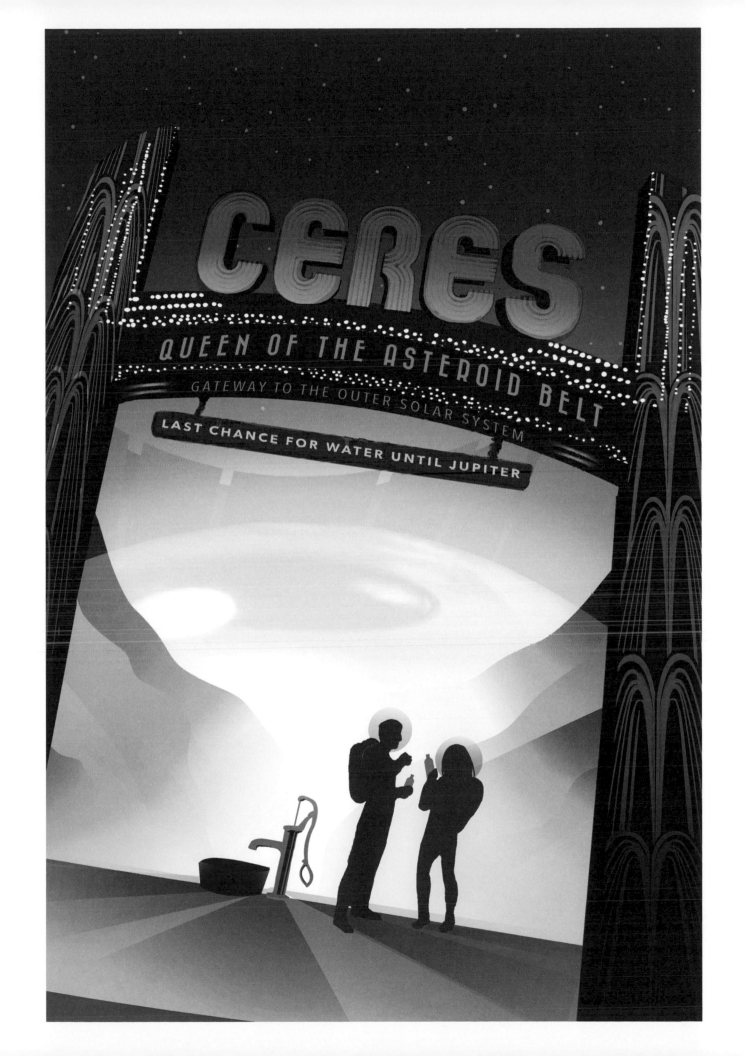

CERES Ceres is the largest asteroid in the solar system. Even though it is small compared to the planets (only about 600 miles [965 kilometers] across), it is still large enough to have segregated its interior into a core, mantle, and crust, so it has been classified as a dwarf planet by astronomers. Ceres is also special because of its icy composition, which future space explorers (and miners) are almost certain to exploit as a rich source of water for drinking, radiation shielding, as well as rocket fuel and breathable air (when splitting it into its constituents hydrogen and oxygen).

Credit: Courtesy NASA/JPL-Caltech

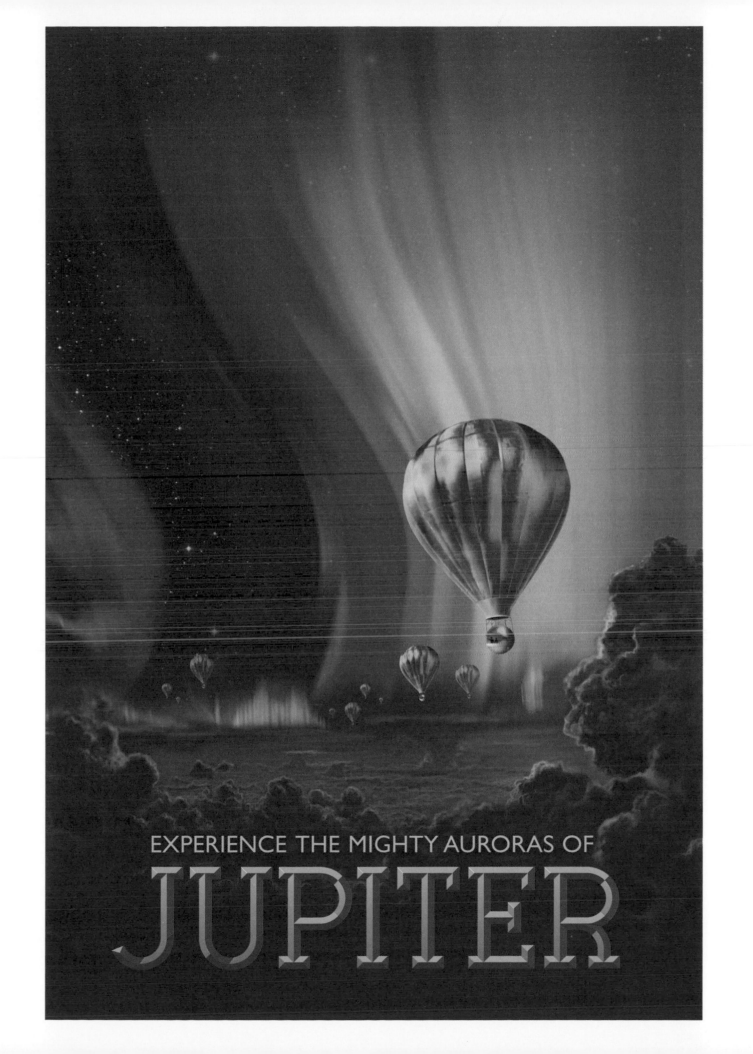

JUPITER The giant planet Jupiter is more than 2.5 times more massive than all of the other planets, moons, asteroids, and comets in the solar system combined, and Jupiter's magnetic field is the largest planetary structure in the solar system. The planet's spectacular polar auroral displays are created by that strong magnetic field, and viewing these beautiful and enormous "northern lights" is sure to be a popular tourist activity in the future.

Credit: Courtesy NASA/JPL-Caltech

EUROPA Europa, the smallest of the large moons of Jupiter discovered by Galileo back in 1610, is covered by a thin, smooth shell of ice that overlies what scientists believe could be the largest salty ocean in the solar system (with perhaps two or three times more water than Earth's oceans). The ocean is in contact with a rocky mantle that is heated by tidal forces from Jupiter and the other large moons. Liquid water, heat sources, and organic molecules are the ingredients required to make a world habitable to life. What lurks in Europa's ocean? Future scientists as well as tourists will be driven to find out.

Credit: Courtesy NASA/JPL-Caltech

TITAN Saturn's largest moon, Titan (which is larger than the planet Mercury), is the only moon in the solar system with a thick atmosphere. The surface pressure on Titan is about 50 percent higher than Earth's, but the temperature is only around -290 degrees Fahrenheit (-180 degrees Celsius), a chilly 90 degrees Celsius above absolute zero. At those temperatures and pressures, the methane, ethane, and other hydrocarbons common in the outer solar system exist as liquids, and they flow across the water-ice surface of Titan in rivers, lakes, and small seas. Titan is similar in some ways to the early Earth before life made our environment oxygen rich. And as the only other place in the solar system besides Earth where liquids flow freely over the surface, it is sure to be a popular and exciting tourist destination in the future.

Credit: Courtesy NASA/JPL-Caltech

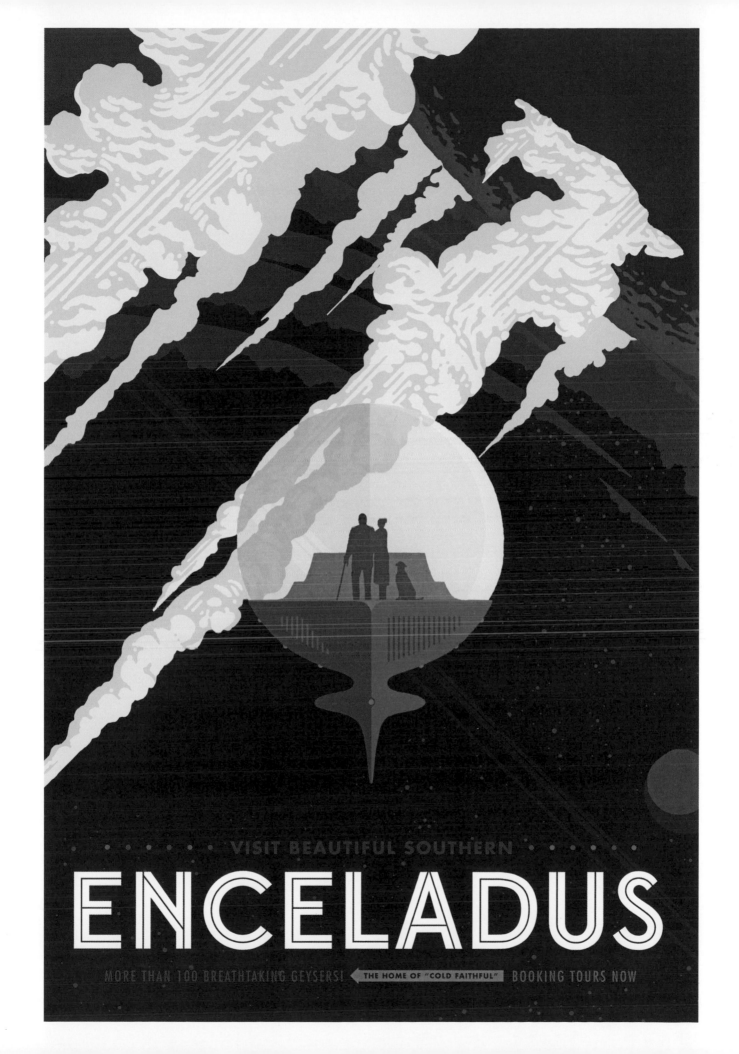

ENCELADUS Tiny Enceladus, only about 310 miles (500 kilometers) in diameter, was long thought to be just another of the many midsize icy moons of Saturn. But when geysers made of water vapor and containing organic molecules were discovered spewing out of that moon's south pole by the NASA *Cassini* mission, Enceladus quickly joined only a few other places in the solar system (like Mars and Europa) where the ingredients for life—liquid water, heat sources, and organic molecules—are known to exist. Visiting those geysers and exploring whatever secrets they might reveal about potential extraterrestrial life is sure to be a popular tourist activity in the future.

Credit: Courtesy NASA/JPL-Caltech

TRAPPIST-1 Using telescopes from Earth and out in space, astronomers have discovered thousands of planets orbiting other Sunlike stars, and more are being discovered all the time. Among the most interesting has been the discovery of seven approximately Earth-size terrestrial planets around the red dwarf star known as TRAPPIST-1. The star is relatively close to the Sun—only 40 light-years away—and so it could very well be a destination for future science missions and even tourist visits.

Credit: Courtesy NASA/JPL-Caltech